Electromagnetic Theory for Complete Idiots

All Rights Reserved. No part of this publication may be reproduced in any form or by any means, including scanning, photocopying, or otherwise without prior written permission of the copyright holder.
Copyright © 2022

Table of Contents

1. VECTOR ANALYSIS

2. VECTOR CALCULUS

3. CURVILINEAR COORDINATE SYSTEMS

4. ELECTROSTATICS

5. ELECTRIC POTENTIAL

6. CONDUCTORS & DIELECTRICS

7. MAGNETOSTATICS

8. MAGNETIC POTENTIAL

9. ELECTROMAGNETIC INDUCTION

10. MAXWELL'S EQUATIONS

APPENDIX

1. VECTOR ANALYSIS

1.1 SCALARS AND VECTORS

In physics, quantities can be classified into 2 main categories; Scalars and Vectors. Certain quantities such as mass, time, temperature, volume, density etc. can be adequately represented just by a numerical value, an amount or a magnitude. Such quantities are called Scalars. For example, if the mass of an object is given as 2 kilograms, the value 2 denotes the number of times the unit kilogram is contained in that object.

On the other hand, certain other quantities such as force, velocity, displacement, acceleration etc. cannot be adequately represented by a numerical value alone. The only way to meaningfully represent such quantities is by considering them as having a direction in addition to the magnitude. Such quantities that have a magnitude as well as a direction are called Vectors. For example, if you walk 2 miles north, then walk 2 miles south, you are back at your original position i.e. your net displacement is zero. Now instead had we just mentioned the magnitude and ignored the directions, there's no way of telling your net displacement, you could have walked 4 miles straight or 4 miles left or whichever way possible.

Vectors are represented by a directed line segment, basically arrows. The length of the line segment denotes the magnitude and the arrowhead indicates the direction of the vector. For example, a vector from point A to another point B is denoted by $\overrightarrow{AB}$. Point A is called the tail of the vector and point B is called the head of the vector. The magnitude of a vector $\overrightarrow{AB}$ is denoted as $|AB|$.

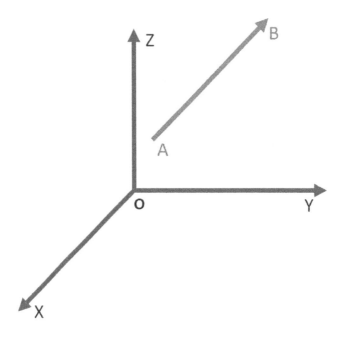

Note that a vector is characterized only by its magnitude and direction, and not by its position in space. This implies that as long as the magnitude and the direction are unaltered, you are free to move the vector around.

1.2 SCALAR MULTIPLICATION

When a vector is multiplied by a positive scalar quantity (a positive number), its magnitude gets multiplied by the scalar quantity and the direction remains unaltered. Similarly, if a vector is multiplied by a negative scalar quantity, its magnitude gets multiplied by the scalar quantity and its direction reverses.

The product of a scalar x and vector $\vec{A}$ is denoted as x$\vec{A}$.

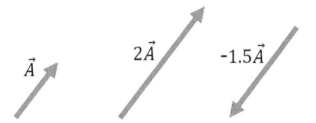

1.3 ADDITION OF VECTORS

Vector addition isn't as straightforward as adding scalar quantities. In case of vectors, the directions have to be considered as well. Consider the following example of a solid block being applied 10 Newtons of force from 2 directions as shown in the figure below.

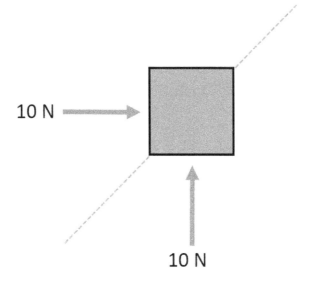

In which direction do you think the block will move as a result of these 2 forces? Of course, along the dotted line. Now imagine that the block is being pushed with 20 Newtons of force from the bottom and 10 Newtons of force from the side as before. In which direction do you think the block will move now? Common sense says the block will move in a direction slightly to the left of the dotted line.

What we just did with these 2 examples is Vector addition. Now let's consider a general case where 2 vectors are oriented at a certain angle from one other as shown below.

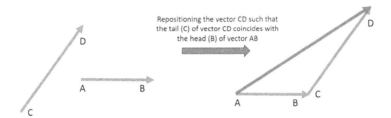

To find the sum or the resultant of 2 vectors, reposition any one of the vectors such that the tail of the one vector coincides with the head of the other vector, then the line segment that completes the triangle denotes the resultant vector. This is called the triangle law of vector addition. In the figure above, we repositioned vector CD such that the tail (C) of vector CD coincides with the head (B) of vector AB. Hence the resultant is given by the vector AD. Now if you try the same, but by repositioning vector AB instead of CD, you will get the exact same result, proving that vector addition is commutative.

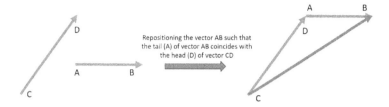

We can extend this idea to any no of vectors and this rule is known as the Polygon law of vector addition.

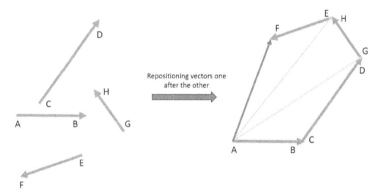

1.4 UNIT VECTOR

Along a particular direction there can be infinite no. of possible vectors with all of them differing in magnitudes from one another. But in any direction, there can only be a single vector of unit magnitude, such a vector is called the Unit vector. The significance of the Unit vector is that all vectors in a specific direction are scaled versions of the unit vector in that direction.

A unit vector in the direction of a vector $\vec{A}$, denoted by $\hat{A}$ can be obtained as,

$$\hat{A} = \frac{\vec{A}}{|A|}$$

1.5 RESOLUTION OF VECTOR

One big advantage of using vectors is that they can be resolved into any no. of components. In the simplest case, a vector can be resolved into 2 component vectors lying in the same plane.

Shown below is a vector $\vec{OA}$, if we draw its projections onto any 2 perpendicular axes in its plane (for convenience we have picked the x and y axes), we get a set of 2 new vectors $\vec{OU}$ and $\vec{OV}$. These 2 new vectors are called the components of vector $\vec{OA}$.

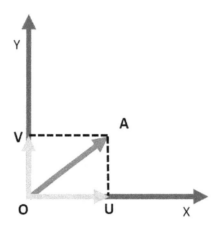

We can take this one step further and resolve a vector into 3 components in 3 dimension. What's the advantage of doing this? For one, resolving vectors into perpendicular components allows us to express them in terms of the Cartesian coordinates, which we are familiar with.

Consider a vector $\vec{OB}$ as shown in the figure below. If we resolve this vector into 2 perpendicular vectors, we get 2 component vectors $\vec{OG}$ and $\vec{OV}$. Vector $\vec{OG}$ is along the z- axis, so we'll leave it as such. But vector $\vec{OV}$ on the other hand lies in the x-y plane, so we to need to resolve it further. Now if we consider $\vec{OV}$ as a separate vector and further resolve it, we get its 2 components $\vec{OE}$ and $\vec{OF}$. This way we have resolved the vector $\vec{OB}$ into 3 mutually perpendicular vectors $\vec{OE}, \vec{OF}$ and $\vec{OG}$ which lies along the x, y and z axes respectively.

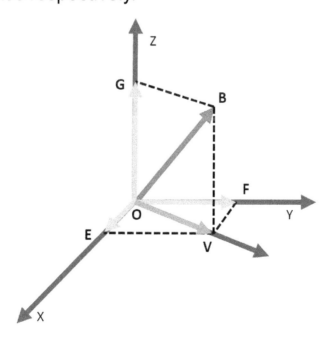

$$\vec{OB} = \vec{OE} + \vec{OF} + \vec{OG}$$

Any vector can be resolved into components along the coordinate axes in this manner. By doing this, we have managed to generalize vectors a little bit. What if we somehow inculcate the idea of unit vectors from last section into this whole thing, Can we generalize vectors even further?

The idea is that the unit vector is the most basic vector along a direction and every other vector along that same direction is just a scaled version of that unit vector. So if we denote the unit vectors along x, y and z axes as $\hat{a}_x$, $\hat{a}_y$ and $\hat{a}_z$ respectively, then any vector A can be expressed as:

$$\vec{A} = \text{(Magnitude in x direction)} \times \hat{a}_x$$
$$+ \text{(Magnitude in y direction)} \times \hat{a}_y$$
$$+ \text{(Magnitude in z direction)} \times \hat{a}_z$$

Math books usually denote the unit vectors along the 3 axes as $\hat{i}$, $\hat{j}$ and $\hat{k}$, but since these notations have other meanings in electrical engineering, the usual practice is to go for $\hat{a}_x$, $\hat{a}_y$ and $\hat{a}_z$.

As a consequence of this generalization, we can denote a vector simply by using 3 no's. For example, (5, 2, -4) denotes the vector $5\hat{a}_x + 2\hat{a}_y - 4\hat{a}_z$. Similarly, (0, 1, 2) denotes the vector $\hat{a}_y + 2\hat{a}_z$. In the later example the x component is absent, meaning the vector lies in the y-z plane.

The magnitude of a vector can be obtained from its components as:

$$\text{Magnitude of a vector} = \sqrt{\begin{array}{l}(\text{Magnitude in x direction})^2 \\ + (\text{Magnitude in y direction})^2 \\ + (\text{Magnitude in z direction})^2\end{array}}$$

This result can be easily obtained by using the Pythagoras theorem.

1.6 DOT PRODUCT

Vectors can be multiplied in 2 possible ways; the Dot product and the Cross product. The dot product of 2 vectors results in a scalar quantity and the cross product of 2 vectors results in another vector. For this reason, the dot product and the cross product are also known as the scalar product and the vector product respectively.

The dot product between 2 vectors is denoted as $\vec{A} \cdot \vec{B}$ (read as A dot B). Mathematically, the dot product is defined as:

$$\vec{A} \cdot \vec{B} = |A||B| \cos \theta$$

$|A|$ = Magnitude of $\vec{A}$
$|B|$ = Magnitude of $\vec{B}$
$\cos \theta$ = cos of the angle between $\vec{A}$ and $\vec{B}$

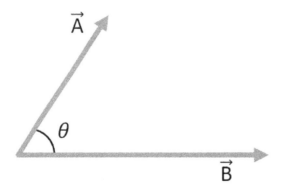

The dot product can be intuitively thought of as a measure of the similarity of two vectors or how well they work together with one another. Consider our "forces on a block" example again.

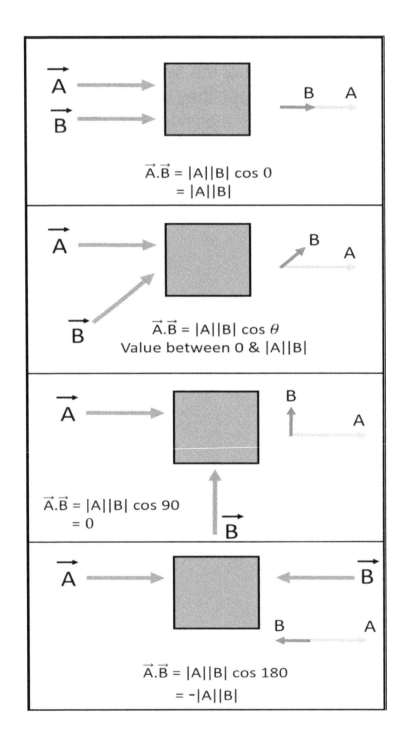

In the first figure, the forces A and B are in the same direction and hence they are working together to move the block. Therefore, the dot product is maximum in this case. In the second figure, the force B is applied at an angle θ to the force A, this obviously is not the best method to move the block. The force B is not contributing to the motion of the block as much as it did in the first case, hence the dot product is a smaller value, but greater than zero. In the third figure, the force B is applied orthogonally to the force A, so it doesn't contribute to moving the block (in the direction of A) at all, hence the dot product here is zero. In the fourth example, the force B is applied in opposite direction to force A, which means force B is not only not contributing to the motion of the block, but it's actually negating the effect of force A. So, the dot product in this case is negative.

Properties of Dot Product:

- Commutative property: **u.v = v.u**

- Distributive property: **u.(v + w) = u.v + u.w**
- **(u + v).(w + z) = u.w + u.z + v.w + v.z**
- **k (u.v) = (ku).v = u.(kv)**, where k is a scalar
- u.u = $|u|^2$
- $a_x \cdot a_x = a_y \cdot a_y = a_z \cdot a_z = 1$
- $a_x \cdot a_y = a_y \cdot a_z = a_z \cdot a_x = 0$

Example:

$\vec{A} = \hat{a}_x + 2\hat{a}_y - 3\hat{a}_z, \vec{B} = 3\hat{a}_x + 5\hat{a}_y + 7\hat{a}_z$

$\vec{A}.\vec{B} = (\hat{a}_x + 2\hat{a}_y - 3\hat{a}_z).(3\hat{a}_x + 5\hat{a}_y + 7\hat{a}_z)$

$\quad = (\hat{a}_x. 3\hat{a}_x + 2\hat{a}_y. 5\hat{a}_y - 3\hat{a}_z. 7\hat{a}_z)$

$\quad = 3 + 10 - 21$

$\quad = -8$

1.7 CROSS PRODUCT

The second way of multiplying 2 vectors is called the Cross product. As mentioned earlier, the cross product of two vectors results in another vector.

Mathematically, the cross product is defined as:

$$\vec{A} \times \vec{B} = |A||B| \sin \theta \, \hat{n}$$

$|A|$ = Magnitude of $\vec{A}$
$|B|$ = Magnitude of $\vec{B}$
$\sin \theta$ = sin of the angle between $\vec{A}$ and $\vec{B}$
$\hat{n}$ = Unit vector perpendicular to both the vectors

Defining the cross product isn't as straightforward as with the dot product. Given 2 vectors, the cross product of the two vectors represents the area of the parallelogram formed with these 2 vectors taken as adjacent sides. And the direction of the cross product vector is normal to plane of this parallelogram. The cross product can be intuitively thought of as a measure of the orthogonality of 2 vectors. Closer the angle between the vectors is to 90 degrees, larger the area of the parallelogram formed and correspondingly larger the cross product vector.

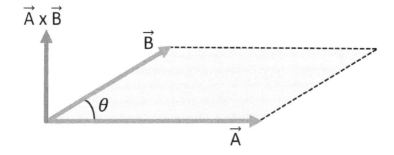

Now there's another small problem, given 2 vectors A and B, there are two possible directions that the cross product vector could point to, upward and downward. This is where the right hand thumb rule comes in, which states that *"If you curl your fingers of your right hand in such a way that the index finger points in the direction of vector A and middle finger points in the direction of vector B, then the thumb points in the direction of AxB".* Now if you try the same for BxA, the thumb will point in the downward direction i.e. BxA = -AxB.

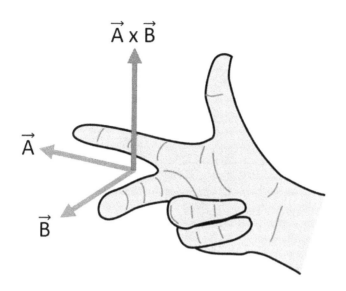

Properties of Cross Product:

- u x v = -v x u
- Distributive property: **u x (v + w) = (u x v) + (u x w)**
- **k (u x v) = (ku) x v = u x (kv)**, where k is a scalar
- u x u = 0
- $a_x \times a_x = a_y \times a_y = a_z \times a_z = 0$
- $a_x \times a_y = a_z$, $a_y \times a_z = a_x$, $a_z \times a_x = a_y$

- $a_y \times a_x = -a_z$, $a_z \times a_y = -a_x$, $a_x \times a_z = -a_y$

Example:

$\vec{A} = \hat{a}_x + 2\hat{a}_y - 3\hat{a}_z$, $\vec{B} = 3\hat{a}_x + 5\hat{a}_y + 7\hat{a}_z$

$\vec{A} \times \vec{B} = (\hat{a}_x + 2\hat{a}_y - 3\hat{a}_z) \times (3\hat{a}_x + 5\hat{a}_y + 7\hat{a}_z)$

$$= \begin{vmatrix} \hat{a}_x & \hat{a}_y & \hat{a}_z \\ 1 & 2 & -3 \\ 3 & 5 & 7 \end{vmatrix}$$

← Vectors can be arranged as following and the Determinant gives the Cross product

$= \hat{a}_x (14 + 15) - \hat{a}_y (7 + 9) + \hat{a}_z (5 - 6)$

$= 29\hat{a}_x - 16\hat{a}_y - \hat{a}_z$

2. VECTOR CALCULUS

2.1 SCALAR AND VECTOR FIELDS

What is a field? A field is a physical quantity that can be specified everywhere in space as a function of position (x, y and z coordinates). And there are basically 2 types of fields: Scalar and vector fields.

A scalar field (or a scalar function) associates a scalar value or a magnitude to every point in space.

Example: Distribution of temperature in a room. If Temperature in a room is given by scalar field $T = xy^2z^3$, then at point (1,1,1) the temperature is 1 unit and at another point (2,3,1), the temperature is 18 units and so on

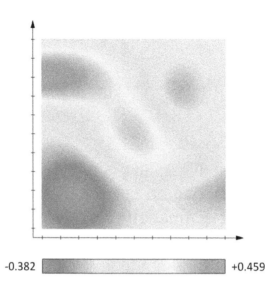

Note that it is not necessary for a scalar field to have non zero values at every point in space.

Similarly, a vector field (or a vector function) is a field in which a vector can be assigned to every point in space.

Example: Velocity of flow at different points in a fluid. If velocity of flow is given by $V = xy\,\hat{a}_x + xy^2z\,\hat{a}_y - z^3\,\hat{a}_z$, then at point (1,1,1) the velocity is denoted by the vector $\hat{a}_x + \hat{a}_y - \hat{a}_z$. At another point

(1,2,3), the velocity is $2\hat{a}_x + 12\hat{a}_y - 27\hat{a}_z$ and so on.

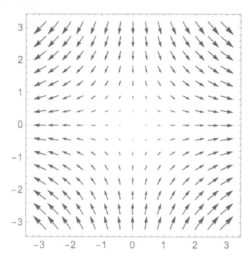

2.2 PARTIAL DERIVATIVE & THE ∇ OPERATOR

For single variable functions, the derivative measures the change in the function value with respect to the independent variable. The partial derivative is the equivalent of the ordinary derivative for multi variable functions. It is basically the derivative of the function with respect to any one of its variables with the other variables held constant. Partial derivative of a function f with respect to x, y & z are denoted as $\frac{\partial f}{\partial x}$, $\frac{\partial f}{\partial y}$ & $\frac{\partial f}{\partial z}$ respectively.

By considering the other variables as constant, we are essentially reducing the multi variable function to a single variable function. And therefore, the partial derivative measures the change in the function value with respect to any one variable.

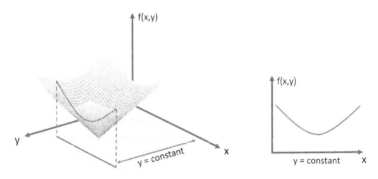

Using partial derivatives, we can define a new operator called the del or the nabla operator, denoted by the symbol ∇. In Cartesian coordinates, the del operator is defined as:

$$\nabla = \frac{\partial}{\partial x}\hat{a}_x + \frac{\partial}{\partial y}\hat{a}_y + \frac{\partial}{\partial z}\hat{a}_z$$

∇ isn't really a vector, it is rather a differential operator. When applied to a single variable function, it is the same as the standard derivative. But for a multi-variable function, it can be applied in 3 different forms (gradient, divergence, curl) and each of these forms has a different significance.

2.3 GRADIENT

A multi variable function has multiple derivatives at every point, one with respect to every variable it depends on. For example, a function **f(x, y)** has 2 derivatives *∂f/∂x* and *∂f/∂y* at every point (shown in the figure below). Each of these derivatives correspond to the rate of change of the function with respect to a particular variable. The resultant or the sum of these derivatives denotes the maximum rate of change of the function when all the variables are considered all at one. That is the gradient.

In other words, the Gradient of a multi variable function is a vector that points in the direction of greatest increase (steepest slope) of the function at a point. It is denoted by symbol ∇f. The gradient is analogous to the slope for single variable functions.

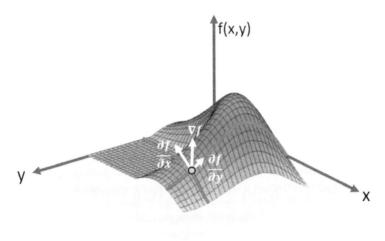

The gradient can be calculated as:

$$\nabla f(x,y,z) = \hat{x}\frac{\partial f}{\partial x} + \hat{y}\frac{\partial f}{\partial y} + \hat{z}\frac{\partial f}{\partial z}$$

Example: f = 2x + yz

$$\nabla f = \frac{\partial (2x + yz)}{\partial x}\hat{a}_x + \frac{\partial (2x + yz)}{\partial y}\hat{a}_y + \frac{\partial (2x + yz)}{\partial z}\hat{a}_z$$

$$= 2\hat{a}_x + z\hat{a}_y + y\hat{a}_z$$

So at a point say P(5,-3, 9), the gradient or the max slope is denoted by the vector $2\hat{a}_x + 9\hat{a}_y - 3\hat{a}_z$.

In another way, the gradient can be thought of as an operator that converts a scalar field (scalar function) into a vector field (vector function).

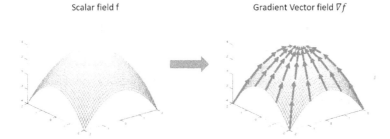

Scalar field f → Gradient Vector field ∇f

Properties of Gradient:

- $\nabla(u + v) = \nabla u + \nabla v$
- $\nabla(uv) = (\nabla u)v + (\nabla v)u$
- $k \nabla u = \nabla(ku)$, where k is a scalar quantity

2.3 DIVERGENCE

We saw how the del operator can be applied on a scalar function in the form of the gradient. On a vector field the del operator can be applied in 2 ways, first of which is the Divergence.

Divergence is nothing but the dot product between the del operator and a vector field.

If $V = V_x \hat{a}_x + V_y \hat{a}_y + V_z \hat{a}_z$,

$\nabla \cdot V = (\frac{\partial}{\partial x} \hat{a}_x + \frac{\partial}{\partial y} \hat{a}_y + \frac{\partial}{\partial z} \hat{a}_z) \cdot (V_x \hat{a}_x + V_y \hat{a}_y + V_z \hat{a}_z)$

$= \frac{\partial V_x}{\partial x} + \frac{\partial V_y}{\partial y} + \frac{\partial V_z}{\partial z}$

Intuitively, the Divergence represents the outward flow of a vector field from an infinitesimal volume at a given point in a vector field. In other words, divergence is a measure of the extent to which a point (which is essentially a tiny volume) behaves as a source of the vector field.

To understand the concept of divergence better, imagine a vector field as a fluid flow as shown below.

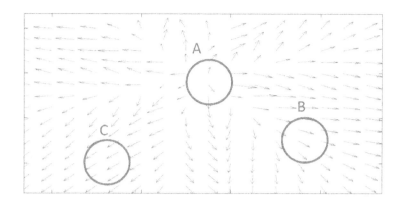

Now if you consider a small spherical volume, the difference between the outward flow and the inward flow i.e. the net outward flow gives the divergence of the flow in the small volume. For example, at point A, all the field lines are pointed away from the volume, which means point A is acting as source of the flux, therefore the divergence at that point is a positive value. At point B, some of the field lines are flowing into the volume and some are flowing out of the volume, but because there are more outward flowing field lines, the net outward flow is positive, therefore divergence is positive at point B as well (but it has less magnitude compared to point A). At point C, there are equal no. of field lines flowing into the volume as there are field lines flowing out of the volume. Hence the divergence at point C is zero.

The above example is only for better understanding of the concept, in reality the divergence has nothing to do with the no. of field lines entering or exiting the volume, it has more to do with the magnitude & direction of the field lines.

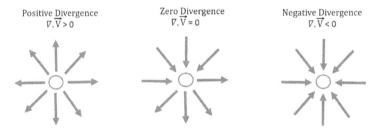

Example: $\vec{V} = x\,\hat{a}_x + yz\,\hat{a}_y + 3xz\,\hat{a}_z$

$\nabla \cdot \vec{V} = (\frac{\partial}{\partial x}\hat{a}_x + \frac{\partial}{\partial y}\hat{a}_y + \frac{\partial}{\partial z}\hat{a}_z) \cdot (x\,\hat{a}_x + yz\,\hat{a}_y + 3xz\,\hat{a}_z)$

$= \frac{\partial(x)}{\partial x} + \frac{\partial(yz)}{\partial y} + \frac{\partial(3xz)}{\partial z}$

$= 1 + z + 3x$

The divergence operator converts a vector field into a scalar field.

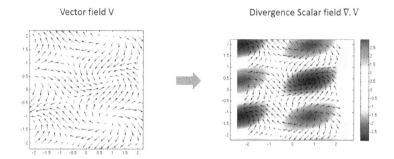

Properties of Divergence:

- $\nabla \cdot (\vec{A} + \vec{B}) = \nabla \cdot \vec{A} + \nabla \cdot \vec{B}$

- $\nabla \cdot (f\vec{A}) = f\nabla \cdot \vec{A} + \vec{A} \cdot \nabla f$, where f is a scalar function

2.4 CURL

The Curl is the cross product between the del operator and a vector field.

If $V = V_x \hat{a}_x + V_y \hat{a}_y + V_z \hat{a}_z$,

$$\nabla \times V = (\frac{\partial}{\partial x} \hat{a}_x + \frac{\partial}{\partial y} \hat{a}_y + \frac{\partial}{\partial z} \hat{a}_z) \times (V_x \hat{a}_x + V_y \hat{a}_y + V_z \hat{a}_z)$$

$$= \begin{vmatrix} \hat{a}_x & \hat{a}_y & \hat{a}_z \\ \frac{\partial}{\partial x} & \frac{\partial}{\partial y} & \frac{\partial}{\partial z} \\ V_x & V_y & V_z \end{vmatrix}$$

$$= (\frac{\partial V_z}{\partial y} - \frac{\partial V_y}{\partial z}) \hat{a}_x - (\frac{\partial V_z}{\partial x} - \frac{\partial V_x}{\partial z}) \hat{a}_y + (\frac{\partial V_y}{\partial x} - \frac{\partial V_x}{\partial y}) \hat{a}_z$$

The curl of a vector field describes the rotational tendency of a vector field at a point in 3d space. Consider a fluid flow as shown in the figure below. Now if you consider a small spherical ball that is free is to rotate in any direction, it will rotate differently depending on its location in the fluid. The field vectors acting on the sphere determines both direction and the speed at which it rotates. Magnitude of the curl vector denotes the speed of rotation and the direction of the curl denotes the axis of rotation of the sphere.

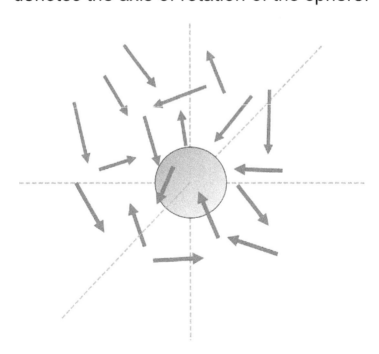

It is important to note that curl refers to the microscopic rotation of the ball at a point in the vector field (i.e. as if translational motion of the ball from the point is restricted) and not the macroscopic circulation of the ball in the field, if any. If you think about the motion of the earth around the sun, the curl is analogous to the rotation of the earth and not its revolution around the sun.

Example: $\vec{V} = x\,\hat{a}_x + yz\,\hat{a}_y + 3xz\,\hat{a}_z$

$$\nabla \times V = (\frac{\partial}{\partial x}\hat{a}_x + \frac{\partial}{\partial y}\hat{a}_y + \frac{\partial}{\partial z}\hat{a}_z) \times (x\,\hat{a}_x + yz\,\hat{a}_y + 3xz\,\hat{a}_z)$$

$$= \begin{vmatrix} \hat{a}_x & \hat{a}_y & \hat{a}_z \\ \frac{\partial}{\partial x} & \frac{\partial}{\partial y} & \frac{\partial}{\partial z} \\ x & yz & 3xz \end{vmatrix}$$

$$= \hat{a}_x \left(\frac{\partial(3xz)}{\partial y} - \frac{\partial(yz)}{\partial z}\right) - \hat{a}_y \left(\frac{\partial(3xz)}{\partial x} - \frac{\partial(x)}{\partial z}\right) + \hat{a}_z \left(\frac{\partial(yz)}{\partial x} - \frac{\partial(x)}{\partial y}\right)$$

$$= -y\,\hat{a}_x - 3z\,\hat{a}_y$$

The curl operator converts a vector field into another vector field.

Properties of Curl:

- $\nabla \times (\vec{A} + \vec{B}) = (\nabla \times \vec{A}) + (\nabla \times \vec{B})$
- $\nabla \times (f\vec{A}) = f(\nabla \times \vec{A}) + \nabla f \times \vec{A}$,where f is a scalar function

2.5 MORE PROPERTIES

- $\nabla \cdot (\nabla \times \vec{A}) = 0$ (i.e. Divergence of Curl = 0)
- $\nabla \times (\nabla f) = 0$ (i.e. Curl of Gradient = 0)
- $\nabla \cdot (\vec{A} \times \vec{B}) = \vec{B} \cdot (\nabla \times \vec{A}) - \vec{A} \cdot (\nabla \times \vec{B})$
- $\nabla \times (\nabla \times \vec{A}) = \nabla(\nabla \cdot \vec{A}) - \nabla^2 \vec{A}$

2.6 LINE INTEGRAL

Line integral or the path integral is the integral of a function evaluated along a curve. For single variable functions, the line integral can only be evaluated over a straight-line path (for obvious reasons) as shown.

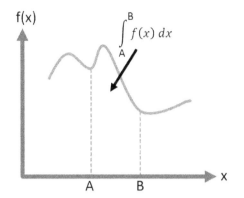 

The line integral is evaluated by considering a bunch of small rectangular strips of width **dx** and suitable height (which will be equal to the value of the function at the point i.e. **f(x)**), so as to fill up the area. The area of an individual strip is therefore **f(x) dx** (width x height) and hence the area of our interest is simply the integral (summation) of these small areas along the path.

For multivariable functions, the line integral can be evaluated along any 2-dimensional path, not just a straight line.

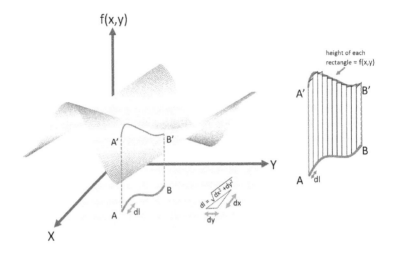

The line integral for multivariable functions is evaluated in the same manner as with single variable functions, except that in this case, the line element (width of the rectangular strip) is no longer **dx** or **dy**, but a function of **dx** and **dy**.

In general, the line integral of a function f along a curve C can be evaluated as:

$$\int_C f \, dl$$

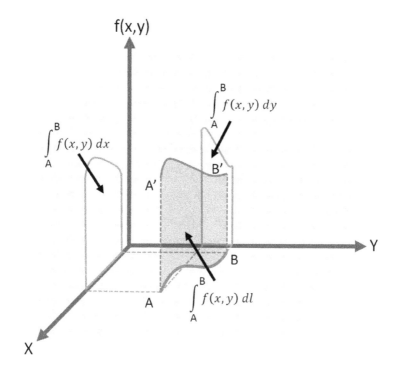

To summarize, the line integral of a scalar function is nothing but the area of the fence created by a curve path and its projection on the function.

Now let's move on to line integration of a vector field. The line integral of a vector field can be interpreted as the amount of work that a force field does on an object as it moves along a curve. When you try to move a block along a curve C in space at constant speed through a force field, a force always acts on the block, making it easier or harder to move the block depending on the directions of the forces at a point. If the force acts opposite to the direction of the path, then you have to do work to keep the block moving. On the other hand, if the direction of the force is in the direction of the path, then your job becomes easier because you're being aided by an outside force.

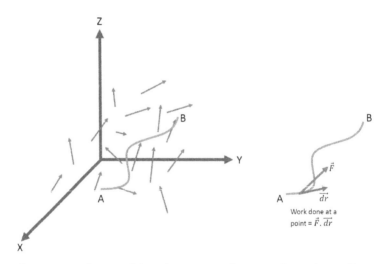

As mentioned in the previous chapter, the dot product is a measure of how well 2 vectors work with each other, hence the dot product between the vector field and the differential curve element integrated along the entire curve gives the total work done.

$\vec{F}$ → Vector field
$\vec{dr}$ → Differential path element

Line integral = $\int_C \vec{F} \cdot \vec{dr}$

2.7 SURFACE INTEGRAL

For scalar functions, the surface integral is nothing but the volume enclosed by a surface and its projection on the function. The simplest case is when the area is planar or 2-dimensional as shown.

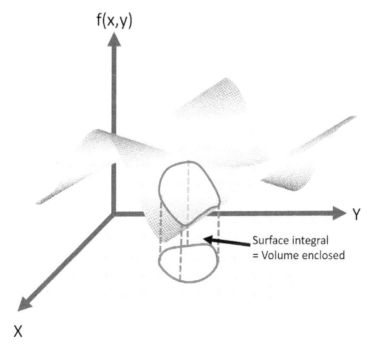

To find the surface integral in such a case, the surface is sliced using planes of small thickness either along the x-axis or the y-axis (along the x-axis in the figure). That way we can convert the surface into a bunch of curves or lines, allowing us to use the line integral.

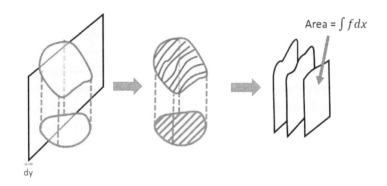

The sliced area is given by the line integral,

$$\int f \, dx$$

Here **dl** =**dx**, since the lines are straight and parallel to the x-axis. To obtain the volume, all we need to do is to combine these areas, which can be done by integrating them with respect to **dy**.

Therefore, the surface integral is given by:

$$\iint f dx\, dy$$

This equation is called the double integral.

But not all surfaces are planar, some surfaces maybe spherical or conical or anything else, in such a case we would require a more generalized method to find the surface integral than the double integral. The idea here is to divide the whole surface into small differential surfaces each of area **dS**, then the volume corresponding to each differential area is the volume of the French fry shaped element (parallelopiped) above it. The volume of this element is **fdS** and hence the volume corresponding to the entire surface is simply the integral of this differential volume.

$$\iint_S f dS$$

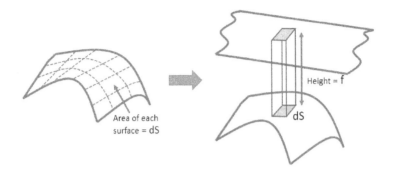

The surface integral of vector field can be interpreted as the amount of field flowing (think of it as fluid flow) through a surface per unit time. When the vector field is normal to the surface, the flow will be maximum and when the surface is parallel to the surface, there will be no flow. So, to calculate the total field flowing through the surface, all we need to do is to add up the component of the vector field that are perpendicular to the surface.

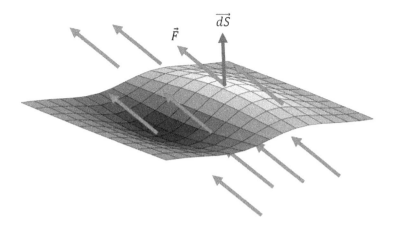

Hence the surface integral of a vector field is given by:

$\vec{F}$ → Vector field
$\overrightarrow{dS}$ → Surface normal vector

$$\text{Surface integral} = \int\int_S \vec{F} \cdot \overrightarrow{dS}$$

2.8 DIVERGENCE THEOREM

According to the Divergence Theorem *"The outward flux of a vector field through a closed surface is equal to the volume integral of the divergence of the vector field over the region enclosed by that closed surface"*.

Mathematically, the Divergence theorem can be written as:

$$\oint_S \vec{F} \cdot \overrightarrow{dS} = \int_V (\nabla \cdot \vec{F}) \, dv$$

It is also known as Gauss's theorem or Ostrogradsky's theorem.

Although the statement and the mathematical formula may look complicated, the intuition behind this theorem is pretty straightforward.

Consider any 3 dimensional object placed a vector field. We have gone with a potato shaped object as shown in the figure. Now let's consider the right hand side of the equation first. We have already learnt that the divergence of a vector field is measure of the outward going or the source like behavior at a point in the field. As this object under our consideration is placed a vector field, every point (small volume) inside the object will have a divergence, either positive or negative or zero. In our figure, we have shown the divergence at 3 points A, B and C. Now if we integrate (sum up) the divergence at all such points throughout the volume, what we get is the total divergence or the total source like nature of the object as a whole, which is essentially how much field originates from the object. This is exactly the same quantity we obtain by taking the surface integral of the field across the full surface.

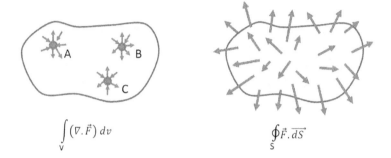

The Divergence theorem in layman's terms can be stated as "*The total outward field flowing out from the surface of an object is the equal to the sum of the fields flowing out from every single point inside the volume of the object*".

What this theorem (and the one in the next section) really does is find a correlation between the smaller phenomenon that occurs on inside of an object to the larger phenomenon on the outer periphery. But this is not the reason we engineers use the divergence theorem, for us the divergence theorem provides a tool for converting surface integrals which are often difficult to compute into an easier volume integral. This is especially useful when we deal with some familiar shape or if the divergence results in a simple function.

An important thing to be noted is that the divergence theorem is only applicable to closed surfaces.

2.8 STOKE'S THEOREM

The Stoke's theorem states that *"the line integral of a vector field round that a closed path is equal to the surface integral of the curl of field over any surface bounded by that closed path"*.

Mathematically, the Stoke's theorem can be written as:

$$\oint_C \vec{F}.\vec{dl} = \int_S (\nabla \times \vec{F}).\vec{dS}$$

The idea behind the Stoke's theorem is quite similar to the divergence theorem. Consider a planar surface placed in a vector field, there will be a rotational effect at every point on the surface which is given by the curl of the field. Now if we integrate (sum up) the curl at all points throughout the surface, what we get is the total rotational effect along the periphery due to the vector field. This is the same as the line integral of the vector field along the outer boundary.

The reason why this is true can be better understood with the help of the figure below. If you drop a ball near to the boundary close to point A, the ball will rotate in clockwise direction due to the curl at the point, given it's acting in isolation. But if you consider the curl at all points on the surface, a ball dropped near point A can no longer keep spinning at A, the curl at point B will inevitably push the ball from point A to point B and then it'll move from point B to point C and so on. In effect the ball dropped at the boundary will only keep moving along the boundary and not the inner surface. The curls on the surface cancel each other out, making it seem like the vector field is just acting along the periphery, not throughout the surface.

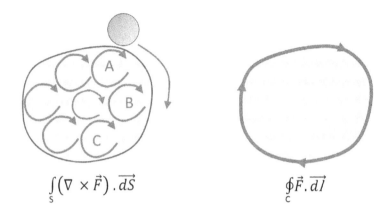

$$\int_S (\nabla \times \vec{F}) \cdot \vec{dS} \qquad \oint_C \vec{F} \cdot \vec{dl}$$

To Vector Calculus in detail, check out our book: Math Beyond Numbers: Vector Analysis

3. CURVILINEAR COORDINATE SYSTEMS

So far in this book, while defining vectors and other stuff, we used cartesian or rectangular coordinate system. While cartesian coordinate system is the simplest and the most popular coordinate system, these are not the most convenient to use when it comes to practical application. In this chapter we'll briefly introduce 2 other coordinate systems: Cylindrical and Spherical. However, for simplicity we'll mostly stick with using cartesian coordinates throughout this book.

3.1 CYLINDRICAL COORDINATE SYSTEM

In Cylindrical coordinate system, a point is space is specified in terms of,

- the distance of the point from the z-axis (ρ)

- the angle a half plane containing the point makes with the x-axis in the anticlockwise direction (θ). This angle is called the azimuth.
- the distance of the point from the xy plane (z) (same as in cartesian coordinates)

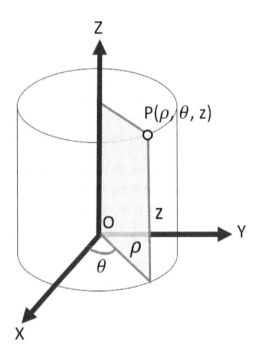

In any coordinate system a point can be defined by the intersection of 3 orthogonal surfaces. In case of cartesian coordinates, these surfaces were 3 planes. For cylindrical coordinate system these surfaces are a cylinder, a plane and a half plane as shown below.

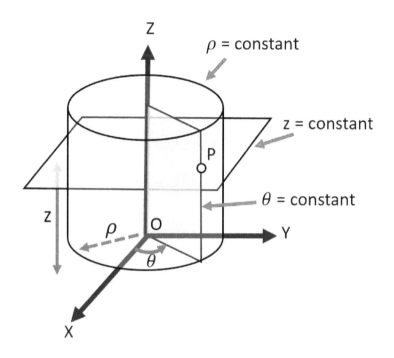

To cover the entire space, the radius of the cylinder ρ can be varied from 0 to ∞, the azimuth angle can be varied from 0 to 2π and the distance from the xy plane can be varied from $-\infty$ to ∞.

Cylindrical to Cartesian coordinates:

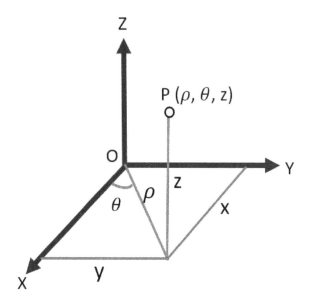

Cylindrical to Cartesian

$x = \rho \cos\theta$ $\rho = \sqrt{x^2 + y^2}$
$y = \rho \sin\theta$ ⇔ $\theta = \tan^{-1}(y/x)$
$z = z$ $z = z$

Vectors in Cylindrical coordinates:

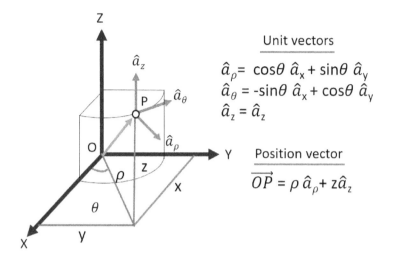

Unit vectors

$\hat{a}_\rho = \cos\theta\, \hat{a}_x + \sin\theta\, \hat{a}_y$
$\hat{a}_\theta = -\sin\theta\, \hat{a}_x + \cos\theta\, \hat{a}_y$
$\hat{a}_z = \hat{a}_z$

Position vector

$\overrightarrow{OP} = \rho\, \hat{a}_\rho + z\hat{a}_z$

3.2 SPHERICAL COORDINATE SYSTEM

In Spherical coordinate system, a point is space is specified in terms of,

- the distance of the point from the origin (r)
- the angle the line joining the point to the origin makes with the z-axis (θ).

- the angle a half plane containing the point makes with the x- axis in the anticlockwise direction (ϕ) (same as the azimuth angle in cylindrical coordinates)

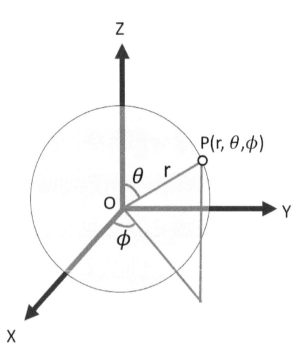

In spherical coordinate system, a point is defined by the intersection of a sphere, a cone and a half plane as shown below.

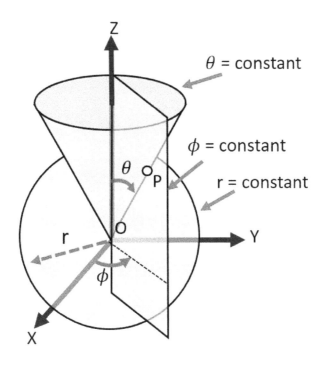

To cover the entire space, the radius of the sphere (r) can be varied from 0 to ∞, the slant angle of the cone (θ) can be varied from 0 to π radians, and the azimuth angle (ϕ) can be varied from 0 to 2π radians.

Spherical to Cartesian coordinates:

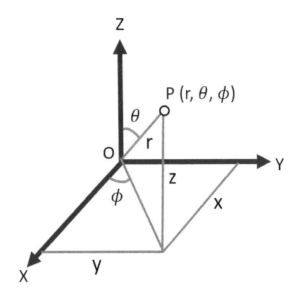

Spherical to Cartesian

$x = r \sin\theta \cos\phi$
$y = r \sin\theta \sin\phi$
$z = r \cos\theta$

$r = \sqrt{x^2 + y^2 + z^2}$
$\theta = \tan^{-1}(\sqrt{x^2 + y^2}/z)$
$\phi = \tan^{-1}(y/x)$

Vectors in Spherical coordinates:

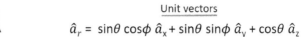

Unit vectors

$\hat{a}_r = \sin\theta \cos\phi \, \hat{a}_x + \sin\theta \sin\phi \, \hat{a}_y + \cos\theta \, \hat{a}_z$

$\hat{a}_\theta = \cos\theta \cos\phi \, \hat{a}_x + \cos\theta \sin\phi \, \hat{a}_y - \sin\theta \, \hat{a}_z$

$\hat{a}_\phi = -\sin\phi \, \hat{a}_x + \cos\phi \, \hat{a}_y$

Position vector

$\overrightarrow{OP} = r\hat{a}_r$

4. ELECTROSTATICS

4.1 COLOUMB'S LAW

Between 1785 and 1787, French engineer & physicist Charles Coulomb performed a series of experiments and established what is now known as Coulomb's law. According to the Coulomb's law "*the force acting between two charges separated by a distance is proportional to the product of the charges and inversely proportional to the square of the distance between them*". And this force acts along the line joining the 2 charges.

If the charges are of same polarity, then the force between them is repulsive and if the charges of opposite polarity, the force between them is attractive.

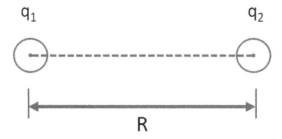

Mathematically, the coulomb's law can be expressed as,

$$F \propto \frac{q_1 q_2}{R^2}$$

$q_1 q_2 \rightarrow$ Product of the charges
$R \rightarrow$ Distance between the charges

To do away with the proportionality sign, Coulomb introduced a constant $k = 1/4\pi\varepsilon$ to the equation, where ε is called the permittivity of the medium. This factor weighs in the effect of the medium in which the charges are located in. For free space, the permittivity is denoted as ε_0 and has a value,

$$\varepsilon_0 = 8.854 \times 10^{-12} = \frac{1}{36\pi} \times 10^{-9} \text{ F/m}$$

For other medium, the permittivity is given by $\varepsilon = \varepsilon_r \varepsilon_0$, where ε_r is the relative permittivity with respect to free space. So the coulomb's law in general can be expressed as,

$$F = \frac{1}{4\pi\varepsilon} \frac{q_1 q_2}{R^2}$$

Just to be clear, when we say the force acting between 2 charges, we mean the force acting on a charge due to the presence of a second charge in vicinity. Similarly, the second charge experiences a force due to the presence of the first charge. These forces are equal in magnitude and opposite in direction in accordance with the newton's 3rd law.

If the 2 charges q_1 and q_2 are located at points having position vectors r_1 and r_2 respectively, then the force on charge q_2 due to charge q_1 can be expressed in vector form as,

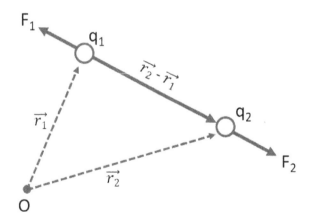

$$\vec{F_2} = \frac{1}{4\pi\varepsilon} \frac{q_1 q_2}{|\vec{r_2} - \vec{r_1}|^2} \hat{a}_{12}$$

Where $\hat{a}_{12}$ = unit vector along $\vec{r_2} - \vec{r_1}$

Further simplification can be made to the formula by expressing the unit vector as,

$$\hat{a}_{12} = \frac{\vec{r_2} - \vec{r_1}}{|\vec{r_2} - \vec{r_1}|}$$

$$\therefore \vec{F_2} = \frac{1}{4\pi\varepsilon} \frac{q_1 q_2}{|\vec{r_2} - \vec{r_1}|^3} \times (\vec{r_2} - \vec{r_1})$$

Obviously the unit vector from q_2 to q_1, a_{21} is the negative of the unit vector a_{12}. Therefore,

$$\vec{F_1} = \frac{1}{4\pi\varepsilon} \frac{q_1 q_2}{|\vec{r_2} - \vec{r_1}|^2} \hat{a}_{21}$$

$$= \frac{1}{4\pi\varepsilon} \frac{q_1 q_2}{|\vec{r_2} - \vec{r_1}|^2} (-\hat{a}_{12})$$

$$= -\vec{F_2}$$

4.2 SUPERPOSITION PRINCIPLE

So far we dealt with the effect of one charge on another, but when there are multiple charges the net effect of all those charges on a single charge can be obtained using the superposition principle. According to the superposition principle, the net force acting on a charge due to N other charges is simply the vector sum of all the forces acting due to each of the N charges taken in isolation.

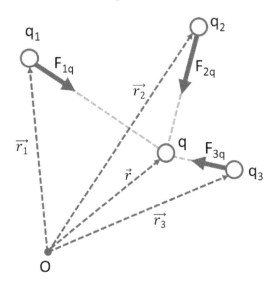

$$\vec{F} = \vec{F}_1 + \vec{F}_2 + \vec{F}_3$$

$$= \frac{q}{4\pi\varepsilon} \frac{q_1}{|\vec{r} - \vec{r}_1|^2} \hat{a}_{q1} + \frac{q}{4\pi\varepsilon} \frac{q_2}{|\vec{r} - \vec{r}_2|^2} \hat{a}_{q2}$$

$$+ \frac{q}{4\pi\varepsilon} \frac{q_3}{|\vec{r} - \vec{r}_3|^2} \hat{a}_{q3}$$

In general,

$$\vec{F} = \frac{q}{4\pi\varepsilon} \sum_{i=1}^{n} \frac{q_i}{|\vec{r} - \vec{r}_i|^2} \hat{a}_{qi}$$

4.3 ELECTRIC FIELD INTENSITY

Every charge has a region around it where the force exerted by it can be experienced by another charge, this region is called the electric field of the charge. For a point charge the electric field is concentric in space. The Electric Field Intensity as its name suggests is a measure of the strength of this electric field. Formally it can be defined as the force experienced by a unit test charge placed in the electric field of another charge.

$$\vec{E} = \vec{F}_{(q_2 = 1)} = \frac{1}{4\pi\varepsilon} \frac{q}{|\vec{r} - \vec{r_p}|^2} \hat{a}_{1p}$$

Where $\vec{r}$ = position vector of charge q
$\vec{r_p}$ = position vector at point P

The Unit of Electric field intensity is Newton/Coulumb.

Just like with forces, the superposition principle is applicable in case of the Electric field intensity too i.e. the Electric field intensity at a point due to multiple charges is equal to the vector sum of the

Electric field intensities due to the charges acting individually.

$$\vec{E} = \frac{1}{4\pi\varepsilon} \sum_{i=1}^{n} \frac{q_i}{|\vec{r} - \vec{r_i}|^2} \hat{a}_{ip}$$

4.4 CONTINUOUS CHARGE DISTRIBUTION

Till now we considered only point charges, but in the real-world charges don't remain concentrated at point, they are distributed over a region. For instance, if you provide charge to a metal rod from one end, the charge doesn't remain concentrated at that end, rather it immediately distributes itself through the length of the rod.

For distributed charges, we use the concept of Charge density. For line elements, the Linear charge density ρ_L (Coulomb/metre) is defined as the total charge per total length of the line. For surfaces, the Surface charge density ρ_S (Coulomb/metre^2) is defined as the total charge per total area of the surface. And for solid objects, the Volume charge density ρ_V (Coulomb/metre³) is defined as the total charge per total volume of the object.

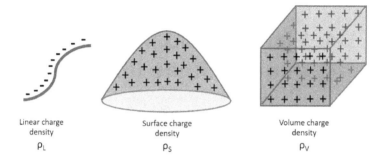

Linear charge density
ρ_L

Surface charge density
ρ_S

Volume charge density
ρ_V

The total charge in each of these cases can be evaluated as:

For Line Charge,

$$Q = \int_L \rho_L \, dl$$

For Surface Charge,

$$Q = \int_S \rho_S \, dS$$

For Volume Charge,

$$Q = \int_V \rho_V \, dV$$

4.5 ELECTRIC FIELD INTENSITY DUE TO DIFFERENT CHARGE DISTRIBUTIONS

1. Electric field intensity due to Infinite line charge:

The electric field intensity due to an infinite line charge can be found by considering the line charge as made up of infinitesimal point charge elements. Consider a tiny line element of length *dl* at a distance R from a point P. If the charge density of the line charge is ρ, then the charge carried by this element is ρ*dl*.

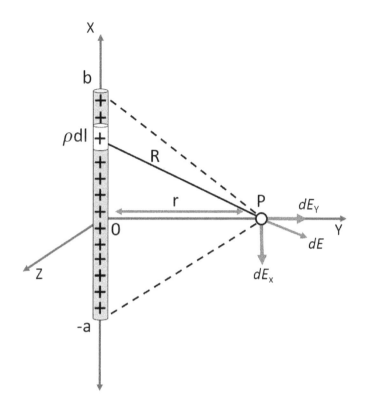

The Electric field intensity at point P due to this element alone is given by,

$$dE = \frac{1}{4\pi\varepsilon} \frac{\rho\, dl}{R^2}$$

Now if you consider an infinite line, then for every line element like the one we considered, there's another such element on the other side of the bisector and therefore the dE_x components cancel each other out and all that's remaining are the dE_y components.

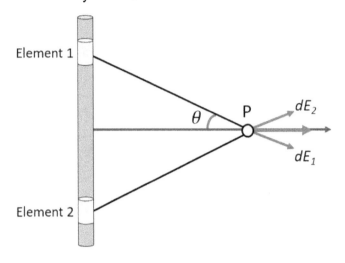

$$\therefore dE = dE_y = dE\cos\theta$$
$$= \frac{1}{4\pi\varepsilon}\frac{\rho\, dl}{R^2}\frac{r}{R}$$

Expressing R in terms of r,

$$dE = \frac{1}{4\pi\varepsilon}\frac{\rho\, dl\, r}{(r^2+l^2)^{3/2}}$$

$$E = \frac{\rho\, r}{4\pi\varepsilon}\int_{-a}^{b}\frac{dl}{(r^2+l^2)^{3/2}}$$

For infinite line charge a = -∞ & b = ∞. Therefore, on solving we get,

$$E = \frac{\rho}{2\pi\,\varepsilon\, r}$$

$$\therefore \vec{E} = \frac{\rho}{2\pi\,\varepsilon\, r}\,\hat{a}_y$$

2. Electric field due to Circular ring

The electric field intensity due to a Circular ring charge can be found in the same way as we did in the previous case, by considering the ring to be made up of infinitesimal point charge elements. Consider a tiny charge element dl on a circular ring of radius r and charge density ρ.

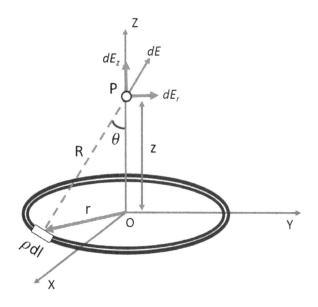

The Electric field intensity at point P at a distance z from the center due to this charge element is,

$$dE = dE_z = dE\cos\theta$$
$$= \frac{1}{4\pi\varepsilon}\frac{\rho\,dl}{R^2}\frac{z}{R}$$

All the radial components cancel each other out due to symmetry and hence only the axial components remain.

$$E = \frac{1}{4\pi\varepsilon} \int \frac{\rho z \, dl}{(r^2+z^2)^{3/2}}$$

$$= \frac{1}{2\varepsilon} \frac{\rho r z}{(r^2+z^2)^{3/2}} \quad [\because \int dl = 2\pi r \text{ (circumference)}]$$

$$\therefore \vec{E} = \frac{1}{2\varepsilon} \frac{\rho r z}{(r^2+z^2)^{3/2}} \hat{a}_z$$

4.6 ELECTRIC FLUX DENSITY

In a previous section, we represented the Electric field using outward pointing vectors (inward for negative charge). But Michael Faraday suggested an alternate way to represent the electric field, using field lines. This provided an easier way to represent more complicated fields.

These field lines have the following properties:

- These field lines originate from a positive charge and terminate at a negative charge.

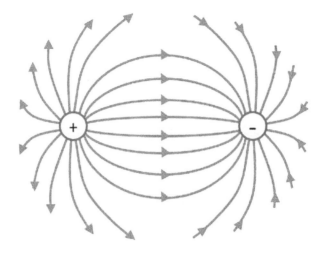

- In the absence of the opposite polarity charge, the field lines from the positive charge terminate at infinity and similarly, the field lines from the negative charge originate at infinity.

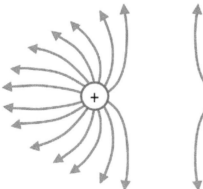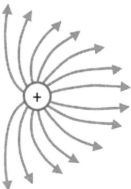

- The more crowed the field lines are, the stronger is the field.
- The field lines never intersect each other.
- The field lines are independent of the medium in which the charge is placed.
- The field lines always enter or exit a surface normally.

As mentioned above, as the number of field lines increases, the field is proportionally stronger. For this new representation, we use the Electric field density denoted by D, as a way to quantify the strength of the Electric field.

So, are both the Electric field density D and the Electric field intensity E the same thing?

They are not! The reason is that the Electric field around a charge is not just depended on the charge, it also depends on the Medium in which the charge is placed. As an Electric field passes through a medium, the atoms tend to polarize adding their own electric field to the existing field (more in chapter 6). The

Electric field Intensity E is the strength of this total field, taking the effect of the medium into account. Whereas, the Electric field density D is just that part of the total electric field discounting the effect of the medium. In free space, the two quantities are related as,

$$E = \frac{D}{\varepsilon_0}$$

4.7 GAUSS'S LAW

The gauss's law states that "**The electric flux passing through any closed surface is equal to the total charge enclosed by that surface**". The idea behind the Gauss's law is straightforward, any flux that flows outward (or inward) from a surface of an object will be the flux generated due to the charge inside the object.

That can't be right! What about the flux due to a charge outside the object? Can't it also pass through the object? These are very valid questions anyone reading about the Gauss's law for the first time should have. To answer these questions, consider a simple surface with an external charge as shown below.

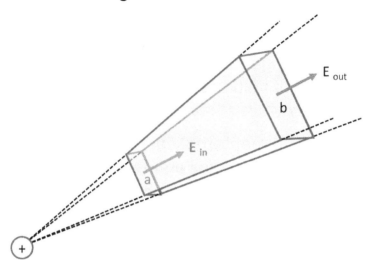

For this surface, more flux will flow out of face **b** than the flux flowing in through face **a** owing to larger area of face **b**. The area of the face and hence the flux flowing, increases by a factor r^2 as the radial distance (r) from the source increases. On the flip side, as r increases the magnitude of the flux decreases by a factor $1/r^2$ as per the inverse square law. This cancels out the increase in the flux due to increase in surface area and takes out the factor r from the equation. Therefore, the net flux though the surface is zero as the flux flowing in through face **a** is equal to the flux flowing out through face **b**. We chose this spherical section for convenience, but this logic works with any shape, however complex it may be. In fact, the gauss's law is applicable to any kind of field that obeys the inverse square law.

To derive a mathematical representation for this law, consider an irregular surface as shown below, with a charge q inside it. This surface can be a real one or an imaginary one, Gauss's law works either way. Now at any point P on the surface, consider a small surface of area dS. The surface vector will obviously be in the direction of the normal at that point, hence the area vector at point P can represented by $\vec{dS}$. Let the flux density at point P be $\vec{D}$. Therefore, the total flux emerging out or passing out of the small surface at P is $\vec{D} \cdot \vec{dS}$.

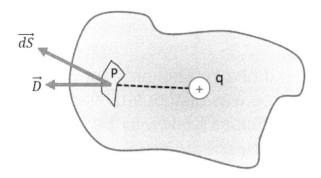

To calculate the total flux emerging out of the surface, all we need to do is to do what we did at point P at every point on the surface i.e. to take the surface integral.

$$\Psi = \oint_S \vec{D}.\vec{dS}$$

Now according to Gauss's law, this quantity is equal to the total charge q enclosed by the surface. Therefore,

$$\oint_S \vec{D}.\vec{dS} = q$$

This is one of the expressions for the Gauss's law. Substituting the Electric field Intensity (E) for the Electric field density (D), we get another expression, perhaps the most popular one.

$$\oint_S \vec{E}.\overrightarrow{dS} = \frac{q}{\varepsilon_0}$$

If instead of a point charge (or charges), the charge was distributed over a volume then the Gauss's law may be written in terms of the charge distribution as,

$$\oint_S \vec{D}.\overrightarrow{dS} = \int_V \rho_V \, dv$$

This equation (in any of these forms) is one of the Maxwell's equations.

4.8 GAUSS'S LAW IN POINT FORM

Another very useful way to express the gauss's law is in the differential form or the point form. The point form can be obtained in the following manner:

$$\oint_S \vec{D}.\vec{dS} = \int_V \rho_v \, dV$$

Using Divergence theorem,

$$\oint_S \vec{D}.\vec{dS} = \int_V (\nabla.\vec{D}) \, dV$$

$$\therefore \int_V (\nabla.\vec{D}) \, dV = \int_V \rho_v \, dV$$

$$\Rightarrow \boxed{\nabla.\vec{D} = \rho_v}$$

Integrating the point form over a volume, we can get back the integral form.

4.9 APPLICATION OF GAUSS'S LAW

The Gauss's law provides a simpler method to find the Electric field intensity for symmetrical charge distributions.

> **1. Electric field intensity due to Infinite line charge:**

We have already derived the Electric field intensity due to Infinite line charge using the direct method. Here we'll try to derive the same using the Gauss's law.

Consider an infinite line charge of linear charge density ρ. Then to calculate the Electric field intensity at a point P at a distance r from the line charge, consider an imaginary gaussian surface passing through this point. In case of a line charge, the gaussian surface is a cylinder.

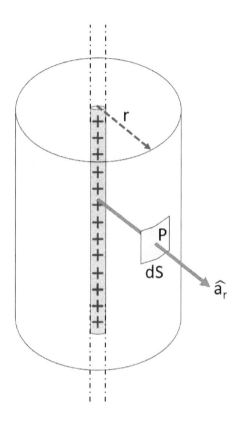

Now, according to the Gauss's law the flux emerging out of the closed surface is equal to the charge enclosed by the surface. Therefore,

$$Q = \oint \vec{D} \cdot \vec{ds}$$
$$= \oint \varepsilon \vec{E} \cdot \vec{ds}$$

Since both the Electric field vector and the surface area vector are both normal to the surface, the dot product is simply the product.

$$\therefore Q = \varepsilon E \oint ds$$
$$= \varepsilon E \times 2\pi r l$$

($\oint ds$ = Surface area of cylinder = $2\pi r l$)

$$\Rightarrow \frac{Q}{l} = \rho = \varepsilon E \times 2\pi r$$

$$\therefore \vec{E} = \frac{\rho}{2\pi \varepsilon r} \hat{a}_r$$

This is exactly the result we obtained through direct method and this derivation is barely two steps. That's the advantage of Gauss's law.

2. Electric field intensity due to Spherical shell:

A shell is a metallic object with a hollow inside. Therefore, the charge resides only on the surface of the shell. Consider a spherical shell of radius **a** with a surface charge density ρ. To find the Electric field intensity at a point at a distance r (from the center of the shell), consider a gaussian surface passing through the point. For a sphere the gaussian surface is another sphere.

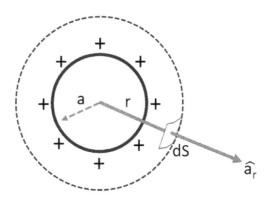

Using Gauss's law,

$$Q = \oint \varepsilon \vec{E} \cdot \vec{ds}$$

$$\Rightarrow Q = \varepsilon E \oint ds$$
$$= \varepsilon E \times 4\pi r^2 \quad (\oint ds = \text{Surface area of sphere} = 4\pi r^2)$$

Surface charge density,
$$\rho = \frac{Q}{4\pi a^2}$$

$$\Rightarrow \rho \times 4\pi a^2 = \varepsilon E \times 4\pi r^2$$

$$\therefore \vec{E} = \frac{\rho}{\varepsilon} \frac{a^2}{r^2} \hat{a}_r$$

In terms of charge Q,

$$\vec{E} = \frac{Q}{4\pi \varepsilon r^2} \hat{a}_r$$

If the point were inside of the shell, then the charge enclosed by the corresponding gaussian surface would be zero and therefore an electric field is never present inside a hollow conductor. This is the principle behind the faraday's cage.

3. Electric field intensity due to Infinite plane sheet:

Consider an infinite plane sheet of charge with surface charge density ρ. Let P be a point at a distance r from the sheet. Now consider a gaussian surface in the form of a cylinder of length 2r perpendicular to the plane sheet. Because we are considering a thin sheet the charge is present on both sides and thus the flux originates from both sides. This is the reason why our gaussian surface extends to both the sides.

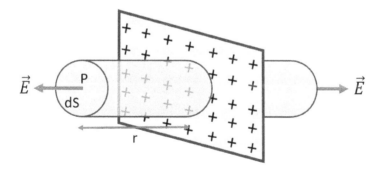

According to Gauss's law,

$$Q = \varepsilon E \oint ds$$
$$= \varepsilon E \times 2 \times A \quad (\oint ds = 2 \times \text{Surface area of cylinder caps} = 2 \times A)$$

Surface charge density,
$$\sigma = \frac{Q}{A}$$

$$\Rightarrow \sigma \times A = \varepsilon E \times 2 \times A$$

$$\therefore E = \frac{\sigma}{2\varepsilon}$$

5. ELECTRIC POTENTIAL

5.1 ELECTRICAL POTENTIAL

Consider 2 positive charges separated by a distance. In order to move any one of these charges close to the other, some external work has to be done, this is to overcome the repulsive force due to the first charge. On the contrary, had one of these charges been positive and the other negative, the charges would have pulled themselves closer without any external work. In that case, the charge is doing the work.

If you remember the explanation for line integral of a vector field, this is the sort of thing we did there too. So, the work done in moving a charge from one point **a** to another point **b** in the presence of any coulombic forces acting on it can be obtained as,

$$W = -\int_a^b \vec{F}.\vec{dl}$$

We can generalize this equation further by considering the work done in moving a unit charge from a point to another. The work done in moving any other charge will simply be a scalar multiple of the work done in moving a unit charge, so this is fair generalization to make.

$$W = -\int_a^b \vec{E}.\vec{dl}$$

How did F turn into E all of a sudden? That's because E is defined as the force acting on a unit charge. Remember??

This difference in the energy levels between 2 points in an Electric field that we have to overcome by doing work in order to move a charge is called the potential difference.

Electric field has an interesting property which we haven't mentioned yet, Electric fields are conservative in nature, meaning the work done in moving charge from one point to another is independent of the path chosen. This implies that moving a charge from point A to B and then from point B to C is the same as directly moving the charge from point A to C. Why is this relevant to our discussion? If we use this idea and pick a common reference point like infinity, then we have a way to define the potential or the energy at a point in an Electric field. Hence the Electric potential(V) can be defined as the work done in moving a unit charge from infinity to a point against an Electric field.

The potential at a point P (r_p) due to a point charge q located at r is given by,

$$V = \frac{1}{4\pi\varepsilon} \frac{q}{|\vec{r} - \vec{r_p}|}$$

One big mistake beginner's make is to confuse the Electrical potential V as a vector function, it is always a scalar function and its unit is Volt.

5.2 CURL OF THE ELECTRIC FIELD

In the last section we touched on something interesting, the conservative nature of the electric field. We said that the work done in moving charge from one point to another is independent of the path chosen. This implies that the work done in moving a charge from a point A to a point B and then from point B back to point A is zero. This is true for any closed path. Mathematically,

$$\oint \vec{E}.\vec{dl} = 0$$

This is essentially the same thing as Kirchhoff voltage law (KVL) in circuit theory, *the algebraic sum of the potential rises and drops around a closed loop is zero.*

Using the Stokes theorem, you can convert the path integral to a surface integral, the surface under consideration being the surface enclosed by this path. Therefore,

$$\oint \vec{E}.\vec{dl} = \int_S (\nabla \times \vec{E}).\vec{dS} = 0$$

$$\Rightarrow \underline{\nabla \times \vec{E} = 0}$$

Now, this is an important result as far as electrostatics is concerned, *the curl of a static Electric field is zero*. It's not hard to figure out this equation from intuition. The electric field produced by a static charge or charge distribution always point away or towards it, so it cannot produce a rotational effect (see section 2.4 for physical interpretation of the curl).

Please bear in mind that this result is not true in general and is applicable only in case of static charges. Only a static Electric field is conservative in nature (more in chapter 9).

5.3 POTENTIAL DUE TO MULTIPLE CHARGES

In the presence of multiple charges, the potential at a point is the work done in moving a unit charge from infinity to the point in the presence of Electric field due to each of these charges. Electrical potential being a scalar quantity, we can simply sum up the potentials due to each charge individually.

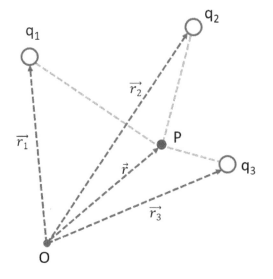

$$V = V_{P1} + V_{P2} + V_{P3}$$

$$= \frac{1}{4\pi\varepsilon} \frac{q_1}{|\vec{r} - \vec{r_1}|} + \frac{1}{4\pi\varepsilon} \frac{q_2}{|\vec{r} - \vec{r_2}|}$$

$$+ \frac{1}{4\pi\varepsilon} \frac{q_3}{|\vec{r} - \vec{r_3}|}$$

In general,

$$V = \frac{1}{4\pi\varepsilon} \sum_{i=1}^{n} \frac{q_i}{|\vec{r} - \vec{r_i}|}$$

5.4 POTENTIAL DUE TO CONTINUOUS CHARGE DISTRIBUTION

For continuous charge distribution the potential at a point P can be obtained by summing up the contributions from individual differential charge elements.

For line charge,
$$V = \int_l \frac{1}{4\pi\varepsilon} \frac{\rho_L \, dl}{|\vec{r} - \vec{r}'|}$$

For surface charge,
$$V = \int_S \frac{1}{4\pi\varepsilon} \frac{\rho_s \, dS}{|\vec{r} - \vec{r}'|}$$

For volume charge,
$$V = \int_V \frac{1}{4\pi\varepsilon} \frac{\rho_v \, dV}{|\vec{r} - \vec{r}'|}$$

$\vec{r}$ = position vector of point P
$\vec{r}'$ = position vector of the element

5.5 EQUIPOTENTIAL SURFACES

An equipotential surface is a surface in which the electrical potential is same throughout. For example, all points at a certain fixed distance from a point charge have the same potential, therefore a sphere surrounding a point charge can be imagined as an equipotential surface. A particular charge distribution can have multiple equipotential surfaces.

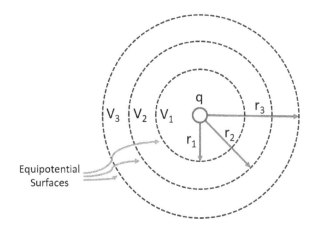

The work done in moving a charge from one point on the equipotential surface to another is zero.

5.6 POTENTIAL GRADIENT

We know how to obtain the Electrical potential function (V) given the Electric field intensity function(E). Now we'll try to do the inverse, i.e. find the Electrical field Intensity function(E) given the potential distribution (V) in space. The challenge here is that the potential is a scalar function and Electrical field Intensity is a vector function. So how would we go from a scalar function to a vector function? Sounds familiar? Ok, let's not jump the gun. There's a pretty formula for this, let's try to derive it.

As a starting point we can use the formulas for point charges and take it from there. The Electrical potential and the Electric field intensity due to a point charge is given by,

$$\vec{E} = \frac{1}{4\pi\varepsilon}\frac{q}{r^2}\hat{r}, \quad V = \frac{1}{4\pi\varepsilon}\frac{q}{r}$$

Where r is the radial distance from the source. Now if we differentiate the Potential function V with respect to r we get,

$$\frac{dV}{dr} = -\frac{1}{4\pi\varepsilon}\frac{q}{r^2}$$

This is almost the Electrical Field intensity function, except for the negative sign and the missing radial unit vector. But that can be taken care of by simply multiplying both sides of the equation by the negative sign and the radial unit vector. Therefore,

$$\vec{E} = -\frac{dV}{dr}\hat{r}$$

Now you must remember that both **V** and **r** are both functions of x, y and z. So we can write this whole thing in terms of the components along x, y and z direction as follows,

$$\vec{E} = -\frac{dV}{dr}\hat{r}$$

$$= -\left(\frac{\partial V}{\partial x}\hat{a}_x + \frac{\partial V}{\partial y}\hat{a}_y + \frac{\partial V}{\partial z}\hat{a}_z\right)$$

$$\therefore \vec{E} = -\nabla V$$

Even without the derivation, we already know that the gradient operator converts a scalar function to a vector function. So there you have it, **the Electric field intensity is the negative gradient of the Electric potential**.

5.7 ELECTRIC DIPOLE

An electric dipole is nothing but a pair of point charges of equal magnitude and opposite polarity separated by a small distance. Electric dipoles deserve special attention simply because dipoles are much more common than isolated charges in the real world.

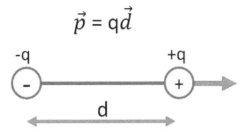

An electric dipole is characterized by the Dipole moment. It is the product of the magnitude of the charge and the distance between them. It is a vector quantity (denoted by letter p) and it is directed from the negative to the positive charge.

The electric field of an electric dipole is the vector sum of the electric fields due to the 2 charges. And the Electrical potential of an electric dipole is the direct sum of the potentials due to the 2 charges.

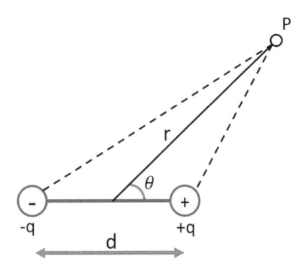

The electrical potential of an electric dipole in terms of the dipole moment is given by,

$$V = \frac{p \cos \theta}{4\pi\varepsilon\, r^2} \quad (r>>d)$$

5.8 POISSON'S AND LAPLACE'S EQUATION

According to the gauss's law,

$$\nabla \cdot \vec{E} = \frac{\rho}{\varepsilon}$$

And from the potential gradient relation we can write Electric field intensity as,

$$\vec{E} = -\nabla V$$

Combining these two results, we can obtain a new relation as follows,

$$\nabla \cdot (-\nabla V) = \frac{\rho}{\varepsilon}$$

$$\Rightarrow \nabla^2 V = \frac{-\rho}{\varepsilon}$$

This relation is called the Poisson's equation. It relates the electrical potential to the charge density which gives rise to it.

As a special case, in a charge free region ($\rho = 0$), this equation simplifies to,

$$\nabla^2 V = 0$$

And it is called the Laplace's equation.

The ∇^2 operator is called the Laplacian. It is the Divergence of the Gradient of a vector field. In cartesian coordinate system,

$$\nabla^2 V = \frac{\partial^2 V}{\partial x^2} + \frac{\partial^2 V}{\partial y^2} + \frac{\partial^2 V}{\partial z^2}$$

The Poisson's Equation when combined with the relevant boundary conditions, can be used to solve for Potential function (V) and thereby Electric field intensity (E) in practical problems. There are several methods to solve the Poisson's and Laplace's equation. The simplest of these is direct integration, which can be applied in case the Electrical potential is a function of one variable.

Example:

Consider a parallel plate capacitor having its plates at **z = 0** and **z = d** with the upper plate at potential V_1 and the lower plate grounded (**V=0**).

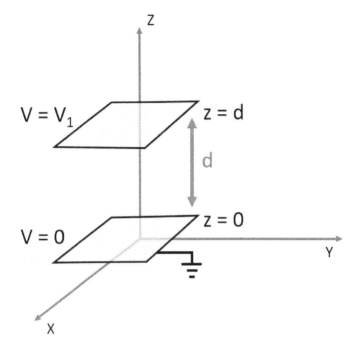

There is no charge density between the plates so we can use the Laplace's equation here. Therefore,

$$\nabla^2 V = 0$$

$$\Rightarrow \frac{\partial^2 V}{\partial z^2} = 0 \quad (\because V \text{ is independent of x and y})$$

On integrating,

$$\frac{\partial V}{\partial z} = C$$

$$V = Cz + D$$

Values of constants C and D can be found using boundary conditions,

at $z = 0$, $V = 0 \Rightarrow D = 0$
at $z = d$, $V = V_1 \Rightarrow C = V_1/d$

$$V = \frac{V_1}{d} z$$

5.9 UNIQUENESS THEOREM

We have seen how useful the Laplace's equation is. Now the question is, whether the Laplace's equation can have multiple solutions? According to the Uniqueness theorem, *the solution to Laplace's equation that satisfies some boundary conditions is unique.* In other words, irrespective of what method you choose to solve the Laplace's equation, the one solution you find will be the only solution.

The way to prove this theorem is by assuming this theorem is false (crazy right?). Assume that we have two solutions of Laplace's equation, V_1 and V_2. Therefore,

$$\nabla^2 V_1 = 0 \quad \& $$
$$\nabla^2 V_2 = 0$$

Now consider a third function $V_3 = V_2 - V_1$. This function V_3 is also a solution to the Laplace's equation because,

$$\nabla^2 V_3 = \nabla^2 (V_2 - V_1)$$
$$= \nabla^2 V_2 - \nabla^2 V_1 = 0$$

Since both V_1 and V_2 are solutions, they must have the same value on the boundary. Therefore, $V_3 = V_2 - V_1 = 0$ on the boundary. The Laplace's equation has an interesting property, that its solution (we have assumed 2 solutions here) cannot have local maxima or minima and its extreme values must occur at the boundaries. Going by this property the maximum and minimum of V_3 are both zero, which implies that V_3 must be zero everywhere and not just the boundaries.

Therefore, $V_1 = V_2$ everywhere and this falsifies our assumption that the Laplace's equation has 2 different solutions. The uniqueness theorem also applies to Poisson's equation and the proof is identical.

6. CONDUCTORS & DIELECTRICS

6.1 CURRENT

We have discussed in length about static charges and the phenomenon associated with them. Now we'll turn our attention to charges in motion. In this chapter we'll take slight detour and study about how different materials behave in an Electric field.

Charges in motion constitute an electric current. An Electric Current can be defined as the rate of flow of charge (electrons). Mathematically,

$$I = \frac{dQ}{dt}$$

The unit of current is Ampere, named after French mathematician and physicist André-Marie Ampère. One ampere of current represents one coulomb of electrical charge moving past a specific point in one second.

Prior to electricity being identified with the electron, it was thought that the movement of positive charges is what produced electric current. For this reason, the direction of current was assumed to be in the direction of flow of positive charges. Such a current is termed as conventional current. Later on, as it was established that the movement of electrons is what is actually responsible for electric current. This current is termed as Electron current and its direction assumed to be opposite to the flow of electrons. For analysis purposes it is easier imagine current as the flow of positive charges.

6.2 CONDUCTORS

Conductors have tons of loosely bound electrons or free electrons inside them and due to thermal energy, they keep moving in random direction. Each of these small movements contribute to an Electric current. But the currents produced by these free electrons are in random direction (opposite to the direction of their motion) and when we consider the conductor as a whole, these currents cancel each other out and the net current is zero.

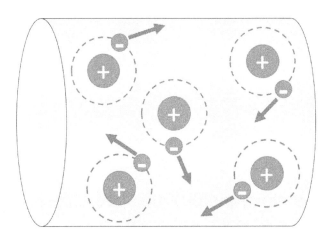

When an electric field is applied, these free electrons experience a coulombic force and as a result the electrons will start moving in the direction opposite to the electric field. But the electrons will still possess the random velocity component due to thermal energy. Because of this the electrons won't parade in a straight line like one might imagine, the randomness just decreases and they sort of move in a common direction.

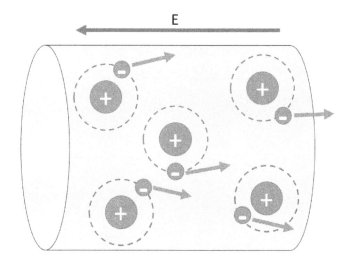

As the electrons pick up speed, they collide with adjacent atoms and the walls of the conductor and as a result they rebound in random direction. Then the electrons once again pick up speed and this process repeats itself. This process is called drifting and the average velocity attained by the electrons during this process is called the drift velocity (v_d).

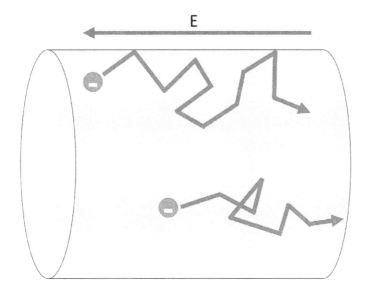

The drift velocity is directly proportional to the Electric field.

$$\vec{v_d} = -\mu \vec{E}$$

Where μ is called the mobility of electrons.

6.3 CURRENT DENSITY

In field theory we are more interested in phenomenon that occur at point rather than within a large region. So it is convenient to introduce a new quantity called current density, denoted by J. The current density at a point is defined as the current that passes through a unit surface area normal to the direction of current. It is a vector quantity, and its unit is A/m². Mathematically,

$$\vec{J} = \frac{dI}{dS} \hat{a}_n \text{ or}$$

$$I = \int_S \vec{J} \cdot \overrightarrow{dS}$$

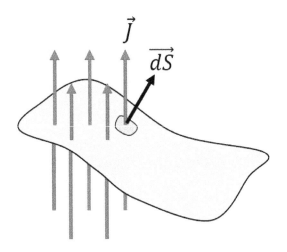

At the particle level, the current density vector indicates the local drift of electrons at a point inside a material and its direction is opposite to the direction of flow of electrons at the point.

Now let's derive the relation between the current density and the drift velocity. Consider a conductor of cross sectional area A as shown in the figure.

Area = A

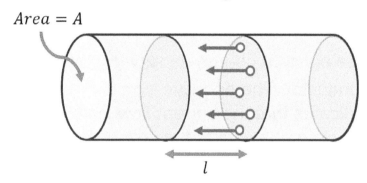

Suppose, in this conductor the electrons are moving with a drift velocity (v_d). Therefore in a small time frame *dt*, the electrons would have travelled a distance *l* = v_d *dt*. In this time period, electrons in the volume given by **V = A l = A v_d dt** would have travelled through the wire. If there are **n** electrons per unit volume, there will be **nV** electrons in the volume V, and the total charge possessed by these electrons will be **dQ = n (A v_d dt) (-e)**.

$$\therefore I = \frac{dQ}{dt} = -neAv_d$$

$$\Rightarrow J = \frac{I}{A} = -nev_d$$

In vector form,

$$\vec{J} = -ne\vec{v_d}$$

Here the quantity (-ne) is nothing but the free electron charge density (ρ_e) of the conductor. The negative sign clearly indicates that the current flow is in the direction opposite to the flow of electrons.

6.4 OHM'S LAW

We are all familiar with the classic version of Ohm's law, V=IR, where V is the voltage, I is the current and R is the resistance. In field theory we can express the ohm's law in a different form, in terms of Current density J and Electric field intensity E.

Consider a conductor of length *l* and cross-sectional area **A** carrying current *I* through it.

$$R = \rho \frac{l}{A} \quad \text{Where, } \rho \rightarrow \text{Resistivity of the material}$$

$$\therefore V = \rho \frac{l}{A} I$$

We know,

$$\frac{I}{A} = J \text{ and } \frac{V}{l} = E$$

$$\Rightarrow J = \sigma E$$

The constant σ is called the Conductivity of the material and its unit is Siemens per meter. For an isotropic conducting medium, the conductivity is the same along all directions, hence this relation can be generalized as,

$$\vec{J} = \sigma \vec{E}$$

This is the vector form or the point form of the Ohm's law.

6.5 CONTINUITY EQUATION

Law of conservation of charge states that **"Electric charge can neither be created nor be destroyed"** and we know that Electric current is nothing but the flow of charges. So from common sense we can infer that any electric current flowing out from a closed surface must be equal to the rate of decrease of the charge enclosed by this closed surface. Mathematically,

$$I = -\frac{dQ}{dt}$$

In terms of current density,

$$I = \int_S \vec{J} \cdot \vec{dS} = -\frac{dQ}{dt} \quad \text{-------} \quad \text{①}$$

Using the divergence theorem,

$$\int_S \vec{J} \cdot \vec{dS} = \int_V (\nabla \cdot \vec{J}) \, dv \quad \text{-------} \quad \text{②}$$

The charge enclosed can be expressed in terms of the volume charge density as,

$$Q = \int_V \rho_v \, dv \quad \text{-------} \quad ③$$

Substituting equation ②,③ in ①

$$\int_V (\nabla \cdot \vec{J}) \, dv = \int_V -\frac{d\rho_v}{dt} \, dv$$

$$\therefore (\nabla \cdot \vec{J}) = -\frac{d\rho_v}{dt}$$

This is the point form of the Continuity equation. Apart from all the mathematical notation this result is a direct consequence of the law of conservation of charge and nothing else. As far as the negative sign is concerned, that's because in this case we considered current as the flow of positive charges. Had we considered electron flow, then the current flowing out would be equal to the rate of increase of electrons within the closed surface, in which case the negative sign wouldn't be required, but the equation will remain the same as the charge density will be negative.

6.6 DIELECTRICS

Unlike conductors, dielectric materials do not possess free electrons. The electrons present in a dielectric are strongly bound to the atom and hence they cannot participate in conduction. But when subject to an Electric field, these bound electrons will experience a coulombic force and this will shift their relative position. In a polarized atom, the center of the electron cloud will no longer align with the center of the nucleus (the proton cloud), forming a dipole of sorts (shown in the figure below). But the atom as a whole will still remain neutral as the no of electrons is equal to the no of protons.

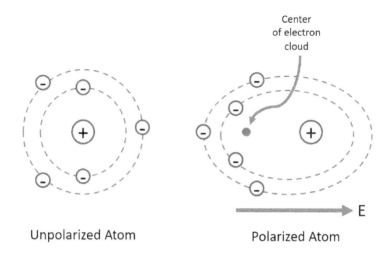

Unpolarized Atom   Polarized Atom

In chapter 4, we mentioned that the Electric flux density(D) is that part of the Electric field intensity(E) that doesn't take the effect of the medium into consideration. The missing link is called Polarization(P). Polarization of a dielectric is defined as the total dipole moment induced per unit volume of a dielectric material. It is a measure of the additional flux density arising due to polarization of a dielectric. Just like Electric flux density D, it is also a vector quantity and has the same unit Coulombs/m².

$$\vec{D} = \varepsilon_0 \vec{E} + \vec{P}$$

The polarization of a dielectric material is directly proportional to the applied electric field. The two are related as,

$$\vec{P} = \varepsilon_0 \chi_0 \vec{E}$$

Where χ_0 is called the electric susceptibility of the dielectric. It tells us how prone or susceptible a dielectric is to polarization in the presence of an electric field. Using the above relation, the Magnetic flux density can be expressed as,

$$\vec{D} = \varepsilon_0 (1 + \chi_0) \vec{E}$$

The factor $(1 + \chi_0)$ is called the dielectric constant or the relative permittivity of a medium with respect to free space.

6.7 BOUNDARY CONDITIONS

Practically, an electric field doesn't always remain confined to free space or a medium. Often it has to pass from one medium to another, it could from free space to a dielectric or from one dielectric to another dielectric, or from a conductor to free space etc. Any way irrespective of the type of media, an Electric field passing between 2 different media satisfy some boundary conditions at the interface. In this section we'll try to establish the general boundary conditions for an electric field at the interface between 2 media.

Consider Electric field passing from one medium to another across a boundary as shown below.

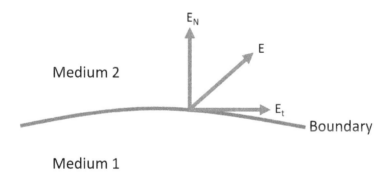

We can resolve the Electric field in both mediums along tangential and normal direction. Then the 2 boundary conditions are obtained by comparing the tangential and the normal components on both sides with each other.

1. Boundary condition for tangential Electric field

To obtain the boundary condition for the tangential component of the Electric field, consider a closed contour **abcd** as shown below.

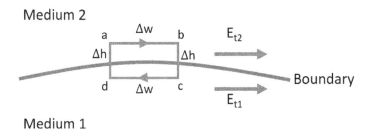

We know that static electric field is conservative in nature i.e. the line integral of the electric field taken along a closed path is zero. Therefore,

$$\oint E.dl = 0$$

$$\Rightarrow \int_a^b E.dl + \int_b^c E.dl + \int_c^d E.dl + \int_d^a E.dl = 0$$

Assume height of the contour Δh to be negligible i.e. $\Delta h \rightarrow 0$

$$\therefore \int_a^b E.dl + \int_c^d E.dl = 0$$

$$\Rightarrow E_{t2} \times \Delta w - E_{t1} \times \Delta w = 0$$

$$\Rightarrow E_{t1} = E_{t2}$$

So the tangential component of the Electric field remains unaltered across an interface between 2 media.

2. Boundary condition for normal Electric field

To obtain the boundary condition for the normal component of the Electric field, consider a gaussian pillbox (just a closed 3d surface) as shown below.

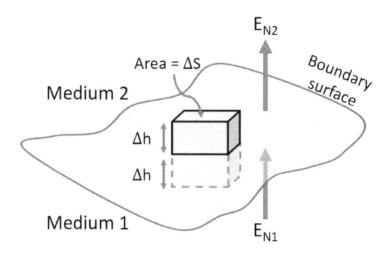

Now we can apply Gauss's law to this Gaussian pillbox as follows,

$$\oint_S E \cdot dS = Q_{enc}$$

Assume height of the gaussian pillbox Δh to be negligible i.e. Δh→ 0. Therefore the flux flows only through top of the box.

$$\therefore \int \varepsilon_2 E_{N2} dS - \int \varepsilon_1 E_{N1} dS = Q_{enc}$$

$$\Rightarrow \varepsilon_2 E_{N2} \times \Delta S - \varepsilon_1 E_{N1} \times \Delta S = Q_{enc}$$

$$\Rightarrow \underline{\underline{\varepsilon_2 E_{N2} - \varepsilon_1 E_{N1} = \sigma}}$$

Where σ is the sheet charge density at the interface. This boundary condition implies that the Normal component of the

Electric field becomes discontinuous as it passes between 2 media.

7. MAGNETOSTATICS

7.1 MAGNETIC FIELD

Static charges produce electric field around them, but charges in motion also produces another kind of field called the magnetic field.

In 1820, Danish physicist Hans Christian Orsted observed that a magnetic field surrounds a current carrying object.

The most fundamental idea in electromagnetism is that there is magnetic field surrounding every current carrying object. These magnetic fields take the shape of concentric rings around a straight wire, called the magnetic field lines. Larger the current flowing through wire, more the no. of magnetic field lines. These lines are not random, they have a direction and it can be determined by using the Right hand thumb rule. It goes like this, *if you point your thumb in the direction of the current, then the fingers curl in the direction of the field lines*

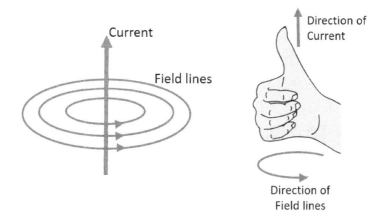

Similarly, when current flows through a coil, a magnetic field is generated around it, such that the coil acts like a magnet with a north and south polarity. The pattern of field lines is as shown below.

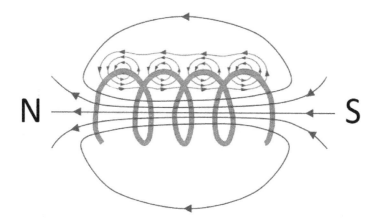

Note that the field lines are concentric if you consider a tiny portion of the coil, but these field lines add and cancel each other giving us this effective pattern. By the way these sorts of coils are called Solenoids.

7.2 MAGNETIC FIELD INTENSITY & MAGNETIC FLUX DENSITY

Just like how the Electric field intensity is a measure of strength of the Electric field, the Magnetic field Intensity denoted as H, is a measure of the strength of a Magnetic field. It is a vector quantity and its unit is Amperes per metre (A/m).

Magnetic flux density (B) on the other hand is the analogous to Electric flux density (D) in electrostatics. It is the total number of magnetic lines of force passing through a unit area. It is also a vector quantity and its unit is Weber per metre square (Wb/m^2) or Tesla.

Both magnetic field intensity & magnetic flux density are measures of strength of magnetic field, but the difference is that the Magnetic flux density depends on the nature of medium.

The two quantities are related to each other as,

$$\vec{B} = \mu \vec{H}$$

Where μ is called the permeability of the medium. It is a measure of the resistance of a material against the formation of a magnetic field. For free space, permeability is denoted as μ_0 and has a value of 4×10^{-7} H/m.

7.3 BIOT SAVART LAW

Biot- Savart Law states that *"the magnetic field intensity produced due to a current element at a point is proportional to the product of the current and the differential length, sine of the angle between the element and the line joining the point to the element, and inversely proportional to the square of the distance between the point and the element"*.

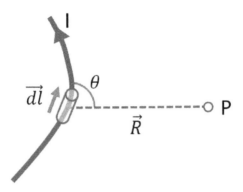

Mathematically, the Biot Savart law can be expressed as,

$$\vec{dH} \propto \frac{Idl \sin\theta}{R^2}$$

dH → Magnetic field Intensity at P due to element dl
dl → Length of the current element
I → Current flowing through the element
R → Distance between P and the element
θ → Angle between the element and the line joining the element and P

The constant of proportionality is k = 1/4π.

In vector form, the Biot Savart law can be expressed as,

$$\vec{dH} = \frac{1}{4\pi} \frac{Idl\ sin\theta}{|\vec{R}|^2} \hat{a}_R$$

7.4 APPLICATION OF BIOT-SAVART LAW

1. Magnetic field intensity due to Infinitely long straight conductor:

Consider an infinitely long straight conductor that carries a current I as shown in the figure. We need to find the Magnetic field intensity at a point P at a perpendicular distance **r** from the conductor. Consider a small current element **dl** at a distance **l** from the origin.

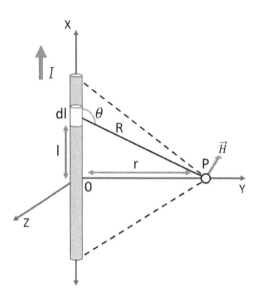

According to Biot-Savart law, the Electric field due to this element *dl* at P is,

$$dH = \frac{1}{4\pi} \frac{I\,dl}{R^2} \sin\theta$$

Expressing R, θ in terms of r,

$$R = \sqrt{r^2 + l^2}, \quad \sin\theta = \frac{r}{\sqrt{r^2+l^2}}$$

$$\therefore dH = \frac{1}{4\pi} \frac{Ir\,dl}{(r^2+l^2)^{3/2}}$$

$$\Rightarrow H = \frac{Ir}{4\pi} \int_{l=-\infty}^{\infty} \frac{dl}{(r^2+l^2)^{3/2}}$$

On integrating we get,

$$H = \frac{I}{4\pi r} \left[\frac{l}{\sqrt{r^2+l^2}} \right]_{-\infty}^{\infty}$$

$$\therefore \vec{H} = \frac{I}{2\pi r} \hat{a}_\phi$$

2. Magnetic field intensity due to a Circular loop:

Consider a circular loop of radius r carrying a current I as shown in the figure.

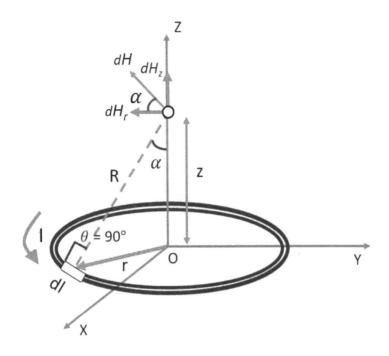

The point P is at a distance **z** from the center of the circular ring. Now if we consider a current element **dl**, then according to Biot-Savart law the Magnetic field intensity due to this element at point P is,

$$dH = \frac{1}{4\pi} \frac{I\,dl}{R^2} \qquad (\theta = 90° \Rightarrow \sin\theta = 1)$$

Due to symmetry, all the cancel out and only axial components remain.

$$\therefore dH = dH_z = dH\sin\alpha$$
$$= \frac{1}{4\pi} \frac{I\,dl}{R^2} \frac{r}{R}$$

Expressing R in terms of r,

$$dH = \frac{1}{4\pi} \frac{Ir\,dl}{(r^2+z^2)^{3/2}}$$

$$\Rightarrow H = \frac{1}{4\pi} \frac{Ir}{(r^2+z^2)^{3/2}} \int dl$$

$$= \frac{1}{4\pi} \frac{Ir}{(r^2+z^2)^{3/2}} \times 2\pi r$$

($\int dl$ = Circumference of loop = $2\pi r$)

$$\therefore \vec{H} = = \frac{Ir^2}{2(r^2+z^2)^{3/2}} \hat{a}_z$$

The Magnetic field Intensity at the center of the Circular ring (z = 0) is given by,

$$\vec{H} = \frac{I}{2r}\hat{a}_z$$

7.5 AMPERE'S CIRCUITAL LAW

From one law to the next. The ampere's circuital law is the magnetostatics version of the Gauss's law for Electrostatics. Ampere's circuital law states that *"the line integral of the Magnetic field intensity H about any closed path is equal to the direct current enclosed by that path"*.

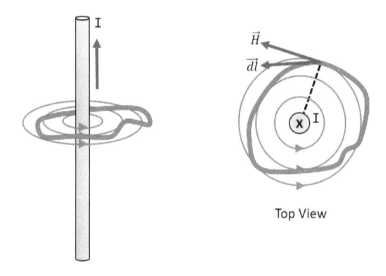

Top View

This closed path can be of any shape, circular, square, or anything random as shown in the figure above.

Mathematically, the Ampere's law can be expressed as,

$$\oint \vec{H} \cdot \vec{dl} = I$$

In terms of the Magnetic field density B, the ampere's law can be expressed as,

$$\oint \vec{B} \cdot \vec{dl} = \mu I$$

Another way to express the Ampere's law is in terms of the current density J.

$$\boxed{\oint_C \vec{B} \cdot \vec{dl} = \mu \int_S \vec{J} \cdot \vec{ds}}$$

In point form, the Ampere's circuital law can be expressed as,

$$\boxed{\nabla \times \vec{B} = \mu \vec{J}}$$

(Derive using Stokes theorem)

It is important to remember that the Ampere's circuital law is only applicable to steady magnetic fields (i.e. those produced by DC currents). This law is the predecessor to one of the Maxwell's equations. The exact Maxwell's equation corresponding to this law, was developed after Maxwell himself modified this law to include the case of varying magnetic fields (more in chapter 10).

7.6 APPLICATION OF THE AMPERE'S CIRCUITAL LAW

The Ampere's circuital law provides an easier method to find the Magnetic field intensity than using the Biot-Savart law in cases where symmetry is present.

1. Electric field intensity due to Infinitely long straight conductor:

Consider an infinitely long straight conductor carrying current I. To find the magnetic field intensity at a point P at a distance **r** from the conductor, consider an amperian path enclosing the conductor passing through the point.

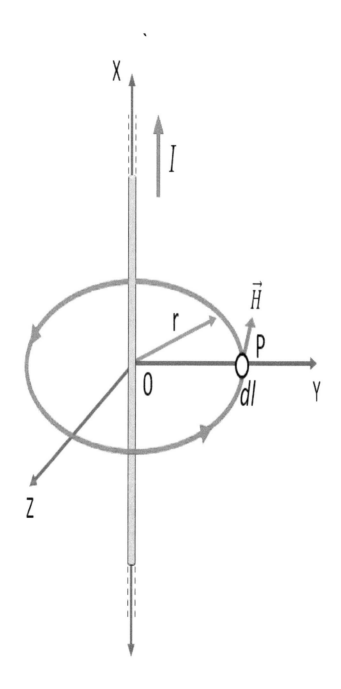

Now if we consider a differential length **dl** at point P, then according to Ampere's circuital law,

$$I = \oint \vec{H} \cdot \vec{dl}$$

Since both the Magnetic field vector and the differential length vector are both tangential to the amperian path, the dot product is simply the product.

$$\therefore I = H \oint \vec{dl}$$
$$= H \times 2\pi r$$

($\oint dl$ = Circumference of loop = $2\pi r$)

$$\therefore \vec{H} = \frac{I}{2\pi r} \hat{a}_\phi$$

2. Electric field intensity due to Infinite plane sheet:

Consider an infinite plane sheet with current flowing as shown in the figure. The surface current density is assumed to be J_s. Now consider an amperian loop of width a as shown.

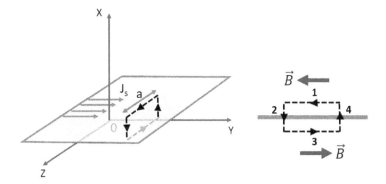

According to Ampere's circuital law,

$$I_{enclosed} = \oint \vec{H} \cdot \vec{dl}$$

Now considering the 4 sides of the amperian loop separately, you can note that there are no field in the direction of sides 2 and 4.

$$\therefore I_{enclosed} = \oint_1 \vec{H} \cdot \vec{dl} + \oint_3 \vec{H} \cdot \vec{dl}$$

$$= 2H \times a \qquad (\oint dl = \text{length of the side} = a)$$

Current enclosed by the path,

$$I_{enclosed} = J_s \times a$$

$$\therefore \vec{H} = \frac{J_s}{2} \hat{a}_n$$

or $$\vec{B} = \frac{\mu J_s}{2} \hat{a}_n$$

Compare this to the equation $E = \rho/ 2\varepsilon_0$ we had for the electric field from an infinite sheet of charge.

7.7 MAGNETIC MONOPOLES

In electricity, there are positive charges and negative charges and they can remain isolated from each other. But in case of Magnetism, there is no such thing as positive pole and a negative pole. Sure, every magnet has a North and a south pole, but they always exist in pairs, i.e. Magnetic monopoles do not exist, at least we haven't found any till date. Not convinced yet? Try breaking a bar magnet into 2 pieces, each new piece will behave as a separate magnet with its own north and south poles. You can do this as much times as you want, but the result will be the same, you will smaller and smaller pieces of magnets, but never a separate north piece and a separate south piece. There are some theory floating around about this behavior of magnets, but exact reason is still largely a mystery.

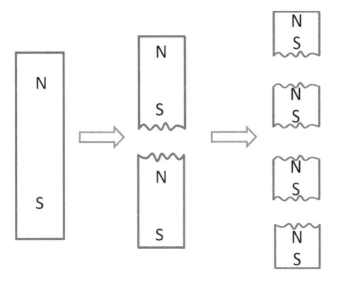

So in reality, there is no source or sink in case of magnets unlike electric fields. The magnetic field lines may seem like they originate at the north pole and terminate at the south pole, but all they are doing is to circulate back on themselves, forming continuous loops.

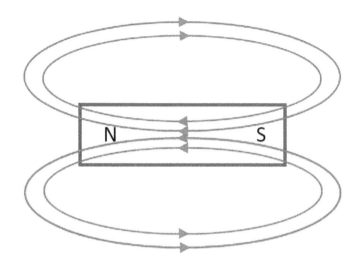

An important inference can be made from this behavior of magnets, that *"The magnetic flux through a closed surface is always zero"*. This result is called the Gauss's law for Magnetism. Connected the dots yet? The reasoning is simple. Because magnetic poles don't exist in isolation, any field lines that enter a volume will inevitably exit from the opposite side, so as to complete the loop, therefore the net flux will be zero.

Mathematically, this law can be put as,

$$\oint_S \vec{B}.\vec{dS} = 0$$

This is one of the 4 Maxwell's equations.

In Vector form, this law can be expressed as,

$$\boxed{\nabla \cdot \vec{B} = 0}$$

8. MAGNETIC POTENTIAL

8.1 MAGNETIC SCALAR & VECTOR POTENTIAL

A potential in the general sense is a quantity you operate upon using an operator to yield a field. In electrostatics, the negative gradient of the Electric potential yielded us the electric field intensity. In case of magnetism, two such potentials can be defined; Magnetic scalar potential & Magnetic vector potential.

The magnetic scalar potential as its name suggests is a scalar quantity. To define the magnetic scalar potential, consider the Ampere's law for a source free region.

$$\nabla \times \vec{H} = 0$$

We know the property of the curl, that the curl of the gradient of a quantity is zero. Using this property, we can express the

Magnetic field intensity H as the gradient of a scalar function. This scalar function is the Magnetic scalar potential, denoted as V_m.

$$\vec{H} = -\nabla V_m$$

In other words, the Magnetic scalar potential can be defined as the scalar function whose negative gradient gives the Magnetic field intensity in a source free region. The negative sign is used by convention. It is analogous to the electric potential.

Always remember that the Magnetic scalar potential is only defined for source free regions.

The magnetic scalar potential like the electric potential also satisfies the Laplace's equation.

$$\nabla^2 V_m = 0$$

To define the magnetic vector potential, we'll start with the Gauss's law for magnetism,

$$\nabla \cdot \vec{B} = 0$$

We know that the divergence of the curl of a vector function is zero. Using this property, we can express the Magnetic flux density B as the curl of a vector function. This vector function is the Magnetic vector potential, denoted as $\vec{A}$.

$$\vec{B} = \nabla \times \vec{A}$$

So the Magnetic vector potential can be defined as the vector function whose curl gives the Magnetic flux density.

The magnetic vector potential does not have a simple physical meaning in the sense that it is not a measurable physical quantity like B or H. But it is a very useful tool as it may be used in regions where the current density is zero or nonzero. Having said that, the Magnetic vector potential is nowhere as useful a tool as Electrical potential was in electrostatics, because when its all said and done it's still a vector function and calculations are still going to be hard (although easier than the Biot-Savart law approach). As far as Magnetic Scalar potential is concerned, its use is limited to source free regions, so that's not much help either.

8.2 LORENTZ'S FORCE EQUATION

A charge **q** moving or stationary placed in an electric field E experiences a force on it. According to Coloumb's law, this force is given by,

$$\vec{F} = q\vec{E}$$

And this force is in the direction of the Electric field E.

On the other hand, a stationary charge placed in a magnetic field doesn't experience any force on it. In a magnetic field, only a charge in motion experiences a magnetic force on it. This force is given by,

$$\vec{F} = q\,\vec{v} \times \vec{B}$$

Where **v** is the velocity with which the charge is moving and B is the Magnetic field intensity. As apparent from the cross product, the direction of this force is perpendicular to both the direction of motion and the direction of the field.

The main difference between the Electric and Magnetic forces is that the force exerted by the electric field is independent of the direction of motion.

So a charge in motion placed in both an Electric and a Magnetic field experiences a combined force given by,

$$\vec{F} = q(\vec{E} + \vec{v} \times \vec{B})$$

This equation is called the Lorentz force law.

Force on a current carrying conductor:

A magnetic force acts on charge moving in a magnetic field, therefore obviously a current carrying conductor, which is basically a system of moving charges, also experiences a force when placed in a magnetic field. The force experienced by a conductor of Length L carrying current I placed in a Magnetic field B can be obtained as,

$$\vec{F} = q\vec{v} \times \vec{B}$$

The velocity vector can expressed as $\vec{v} = \vec{l}/t$, where $\vec{l}$ is the length of the conductor (which is the distance the electron travels).

$$\therefore \vec{F} = \frac{q}{t}(\vec{l} \times \vec{B})$$

$$\Rightarrow \vec{F} = I\vec{L} \times \vec{B}$$
$$= IBL\sin\theta$$

Where θ is the angle between the direction of current (I) and the direction of magnetic field (B).

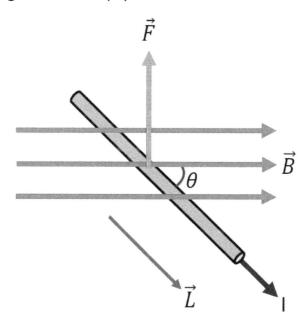

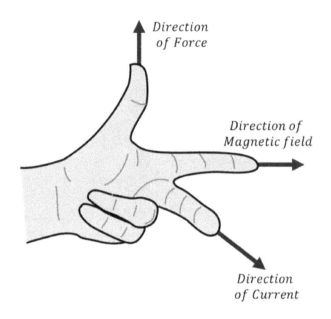

Fleming's Left hand rule

The direction of the force on the wire can be determined using the Fleming's Left Hand Rule, which is stated as; *"Hold the forefinger, middle finger and thumb of your left hand perpendicular to each other. If the forefinger represents the direction of the field and the middle finger represents that of the current, then the thumb points in the direction of the force."*

8.3 MAGNETIC DIPOLE

Similar to an Electric dipole, there exists Magnetic dipoles too. A Magnetic dipole is nothing but an arrangement of 2 magnetic poles separated by a small distance. And the Magnetic Dipole moment or simply magnetic moment is the product of the strength of either pole and the distance between the poles. It is a vector quantity (denoted by letter m) and it is direction is from the south pole to the north pole.

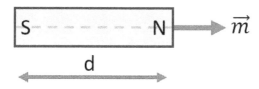

A bar magnet is the simplest example of a magnetic dipole. Another example of a Magnetic dipole is a current carrying loop. A bit odd isn't it? A current carrying loop looks nothing like a dipole! While that's true, the distribution of magnetic lines of force around a finite current carrying loop is similar to that produced by a bar magnet and for this reason a current carrying coil can be considered as a magnetic dipole.

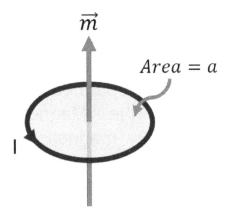

The magnetic dipole moment in this case is given by,

$$\vec{m} = I\vec{a}$$

8.4 MAGNETIC MATERIALS

A magnetic material is a substance that can be magnetized in the presence of an external magnetic field. In this section, we will discuss the nature of magnetic materials. But we will keep the discussion to the basics, since a detailed study will require you to delve into quantum mechanics, which is beyond the scope of this book.

A material is said to be a magnetic material if it possesses dipole moments within it. In general, magnetic dipole moments in an atom can come from 3 sources. First is the orbital motion of the electrons around the nucleus. An electron moving in an orbit is basically the same as a current carrying loop and therefore this motion contributes to a magnetic moment.

The second source of magnetic moment is the spin of the electrons. The electron spins from unpaired electrons can contribute to net magnetic moment. The third source is the nuclear spin. Similar to the electron spin, nuclear spin also contributes to a magnetic moment. But its contribution is negligible, so we'll stick with the other two for this discussion. The total magnetic moments possessed by the atoms of a magnetic material will be the vector sum of these 3 magnetic moments.

Based on the nature of these magnetic moments, we can classify magnetic materials into 3 types (actually there are more, but these are the main types):

Paramagnetic, Diamagnetic and Ferromagnetic materials.

Diamagnetic Material:

In a diamagnetic material, the orbital magnetic moment and the electron spin magnetic moment are equal opposing and they cancel out. Therefore, these materials possess no magnetic moments on their own. However, in the presence of an external magnetic field, the spin magnetic moment increases slightly compared to the orbital magnetic moment and a net magnetic moment is created. This induced magnetic moment will be in the direction opposite to the applied magnetic field. The strength of this induced magnetic moment will be very weak.

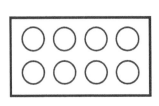

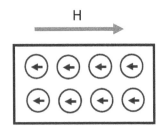

a) Without external Magnetic field   b) In presence of external Magnetic field

Since the polarity of the induced magnetic field is opposite to that of the external field, a diamagnetic material tends to experience a torque that aligns it perpendicular to the external field.

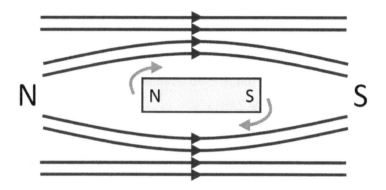

A diamagnetic material does not retain its magnetic properties, after the external field is removed. Common examples of diamagnetic substances are Copper, Zinc, Bismuth, Silver, Gold, Water, Glass, NACL, etc.

Paramagnetic Material:

Contrary to a diamagnetic material, the atoms of a paramagnetic material do have a net magnetic moment. This net magnetic moment is small in magnitude, but non-zero. If paramagnetic materials do possess a magnetic moment, then they should exhibit magnetic properties even without an external field. Right?? The problem is that even though the individual atoms possess a magnetic moment, when the material is considered as a whole, the moments due to the atoms aligned in random direction and therefore the total magnetic moment for the material is zero.

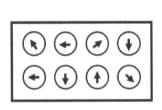 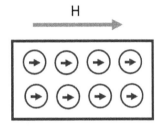

a) Without external Magnetic field   b) In presence of external Magnetic field

The polarity of the induced magnetic field in a paramagnetic material is in same direction as the external field, therefore a paramagnetic material tends to align parallelly to the external field.

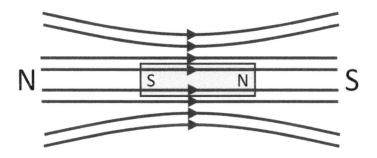

Like the diamagnetic materials, paramagnetic materials also can't retain magnetic properties once the external is removed. Common examples of paramagnetic substances include aluminum, chromium, potassium, oxygen, tungsten etc.

Ferromagnetic Material:

In case of ferromagnetic materials, things are a little different, their atoms possess a relatively strong magnetic moment and they aren't aligned randomly. But due to interaction between neighboring atoms, a bunch of these moments line up into a region called domain. Each of these domains have a strong magnetic moment and all of them vary in direction. Therefore the overall effect is zero. However, in the presence of a magnetic field all of these domains tend to align in the direction of the external field and hence they exhibit strong magnetic properties.

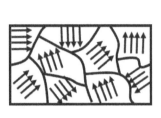

a) Without external Magnetic field

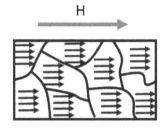

b) In presence of external Magnetic field

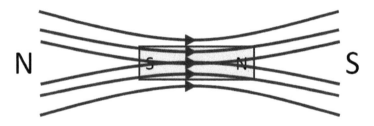

The best part about ferromagnets is that it can retain its magnetic moment (to certain extent) even after the external magnetic field is removed. This is why ferromagnetic material are the preferred choice to make permanent magnets. The only way a ferromagnet loses its magnetic properties once its magnetized is through application of reverse field or through heating. Examples of ferromagnetic materials include iron, nickel, cobalt and a handful of alloys.

8.5 MAGNETIZATION

In the previous section we saw how different materials react to magnetic fields. In this section, we'll introduce a new quantity called Magnetization(M) to quantify all that we learnt in the last section. Magnetization of a magnetic material is defined as the total magnetic dipole moment induced per unit volume of the material. It is a measure of the additional magnetic flux density arising due to creation of magnetic moment in a material. It is a vector quantity.

$$\vec{B} = \mu_0(\vec{H} + \vec{M})$$

The Magnetization of a material is directly proportional to the applied magnetic field. The two are related as,

$$\vec{M} = \chi_m \vec{H}$$

Where X_m is called the magnetic susceptibility of the material. It tells us how prone or susceptible a material is to magnetization in the presence of a magnetic field. Using the above relation, the Magnetic flux density can be expressed as,

$$\vec{B} = \mu_0(1+ X_m)\vec{H}$$

The factor $(1+ X_m)$ is called the relative permittivity of a medium with respect to free space.

Diamagnetic materials have a negative magnetic susceptibility X_m. Whereas both paramagnetic and ferromagnetic materials have a positive magnetic susceptibility, but for ferromagnetic materials $X_m >> 0$.

8.6 BOUNDARY CONDITIONS

Just as the electric field satisfies some boundary conditions, the magnetic field also satisfies certain conditions along the boundary of two different materials. These conditions can be obtained in pretty much the same manner as we did with the Electric field, by resolving the field into tangential and normal components.

1. Boundary condition for tangential Magnetic field

To obtain the boundary condition for the tangential component of the Magnetic field, consider a closed contour **abcd** as shown below.

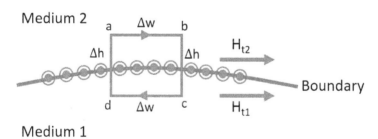

Now, using the Ampere's circuital law we can write,

$$\oint H \cdot dl = I_{enc}$$

$$\Rightarrow \int_a^b H \cdot dl + \int_b^c H \cdot dl + \int_c^d H \cdot dl + \int_d^a H \cdot dl = I_{enc}$$

Assume height of the contour Δh to be negligible i.e. $\Delta h \rightarrow 0$

$$\therefore \int_a^b H \cdot dl + \int_c^d H \cdot dl = I_{enc}$$

$$\Rightarrow H_{t2} \times \Delta w - H_{t1} \times \Delta w = I_{enc}$$

$$\Rightarrow H_{t2} - H_{t1} = I_{enc} / \Delta w$$

$$\underline{\underline{H_{t2} - H_{t1} = K}}$$

The vector quantity K is called the surface current density on the interface. So the tangential component of the Magnetic field unlike that of the Electric field becomes discontinuous as it passes between 2 media.

2. Boundary condition for normal Magnetic field

To obtain the boundary condition for the normal component of the Magnetic field, consider a gaussian pillbox as shown below.

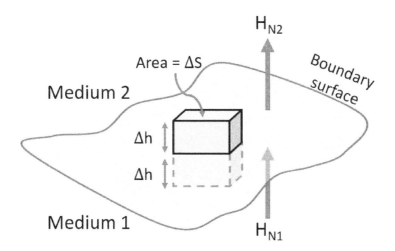

Now, we can apply the gauss's law for magnetism over this gaussian pill as follows,

$$\oint_S H.dS = 0$$

Assume height of the gaussian pillbox Δh to be negligible i.e. Δh⟶ 0. Therefore the flux flows only through top of the box.

$$\therefore \int H_{N2} dS - \int H_{N1} dS = 0$$

$$\Rightarrow H_{N2} \times \Delta S - H_{N1} \times \Delta S = 0$$

$$\Rightarrow H_{N2} - H_{N1} = 0$$

So the normal component of the Magnetic field remains unaltered across an interface between 2 media. These are the general boundary conditions for the Magnetic field across 2 media and depending on the type of the 2 media, these conditions can be modified accordingly.

9. ELECTROMAGNETIC INDUCTION

9.1 FARADAY'S LAWS

In magnetostatics, we dealt with magnetic fields produced by steady currents. Here we'll deal with currents that vary with time.

In 1831, Michael Faraday literally played around with a bunch of bar magnets and some coils, and in the process came up with some laws. No big deal! In one of his experiments, he moved around a bar magnet through a coil connected to an ammeter. He observed that the ammeter showed deflection.

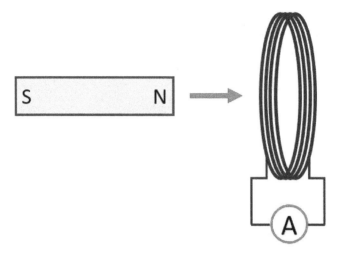

In another experiment, he placed 2 coils close to each other, one connected to a varying current source and the other to an ammeter. What do you think happened there? Again, the ammeter showed some deflection.

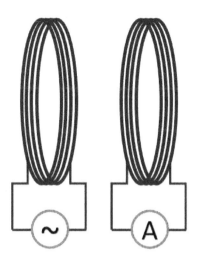

Based on these experiments Michael Faraday formulated 2 laws, which is now known as the Faraday's Laws. These laws introduce us to the phenomenon called Electromagnetic Induction.

According to the Faraday's first law, *when a conductor is placed in a varying magnetic field, an EMF gets induced across the conductor and if the conductor offers a closed circuit then an induced current flows through it.*

And Faraday's second law states that, *the induced EMF is directly proportional to the rate of change of magnetic flux.*

$$\text{emf } \alpha \frac{d\phi}{dt}$$

The important thing to note here is that the magnetic field must be varying. A stationary magnetic field cannot induce a magnetic field. Which brings us to our next question, how do we obtain a varying magnetic field?

There are two ways to obtain varying magnetic field:

1. One is relative spatial movement, that is if the distance between the magnet and the conductor keeps changing, the magnetic field also keeps changing and induction is possible. (as in faraday's first experiment)

2. The other is to vary the magnetic field originating from the source itself. This is not possible with permanent magnets, but it's easy to do using current carrying coils. All we need to do is to vary the current through the coils, the magnetic field also varies as a result. (as in faraday's second experiment)

Now, here's a thought. If the second experiment works, then can we just place many coils in the proximity of a current carrying coil and induce current in all of them? Yes, that's possible. Wait! Did we just invent a new method to generate electricity?? Unfortunately not, there's a catch in all this, called mutual induction. When we induce a current in the secondary coil, this current will itself produce a flux in the secondary coil. This flux will link with primary coil inducing an EMF and a current in the primary, this is a mutual process.

The problem is that the current induced back in the primary will be in the opposite direction to the original applied current in the primary, thus reducing the overall effect. This is a direct consequence of the law of conservation of energy. In electromagnetics, it's called the Lenz's law. The Lenz's law states that "**the direction of the electric current which is induced in a conductor by a changing magnetic field is such that the magnetic field created by the induced current opposes the cause** *that induced it*". Lenz's Law ensures that the electrical energy of the primary coil is reduced by the same amount as the energy gained by the secondary coil. In other words, an induced effect is always such as to oppose the cause that produced it.

For clarity consider the figure below. In this case the north pole of the bar magnet is moving towards the coil and this induces an emf and a current in the coil. To conserve energy, the coil must oppose this motion of the bar magnet and this can be done if the left side of the coil acts as a north pole (north- north repulsion). Therefore, the current is induced in the coil in such a way that a north pole is created on the left side of the coil. If instead the north pole of the bar magnet was pulled away from the coil, then to resist this motion the coil would need to create a south pole on its left side in order to attract the north side of the bar magnet and in turn stop it from moving away. So the current in the coil is induced accordingly.

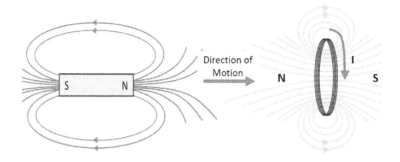

Taking the lenz's law into account, our equation for the induced emf can be modified as,

$$\text{emf} = -N\frac{d\phi}{dt}$$

Here N denotes the no of turns in the coil.

Electromagnetic induction is the principle behind the working of devices like transformers, motors etc.

9.2 INTEGRAL & POINT FORM OF FARADAY'S LAW

We know that the Voltage or the emf can be expressed in terms of the Electric field as,

$$\text{emf} = \oint \vec{E}.\vec{dl}$$

$$\therefore \oint \vec{E}.\vec{dl} = -\frac{d\phi}{dt} \quad \text{------ ①}$$

The magnetic flux can be expressed in terms of Magnetic flux density as,

$$\phi = \int_S \vec{B}.\vec{dS}$$

Therefore equation ① becomes,

$$\boxed{\oint \vec{E}.\vec{dl} = -\frac{d}{dt}\left(\int_S \vec{B}.\vec{dS}\right)}$$

This equation is the integral form of the Faraday's law, and is the final Maxwell's equation. Wait!! There must be a mistake here. In chapter 5 (section 2), we said that electric field is conservative and therefore the line integral around a closed path is zero. And we obtained a related result that the curl of an electric field is zero. Before you start scratching your head over this, it's not a mistake. If you remember clearly, even in that section we mentioned that the conservative behaviour is applicable strictly to static electric fields. Whenever there is a varying magnetic flux, the electric field associated is no longer conservative and the line integral of E.dl over a closed path becomes equal to the EMF. This is the reason why the emf induced in coil depends on the no of turns of the coil as well. Each turn contributes to the line integral and therefore the EMF.

This equation in the point form can be obtained as follows,

Using the Stoke's theorem, the path integral can be expressed as the surface integral.

$$\oint \vec{E}.\vec{dl} = \int_S (\nabla \times \vec{E}).\vec{dS}$$

$$\therefore \int_S (\nabla \times \vec{E}).\vec{dS} = -\int_S \frac{d(\vec{B})}{dt}.\vec{dS}$$

$$\Rightarrow \boxed{\nabla \times \vec{E} = -\frac{d(\vec{B})}{dt}}$$

Unlike the Electric field due to static charges, electric fields induced due to a varying magnetic field is circulating in nature and therefore has a non zero curl.

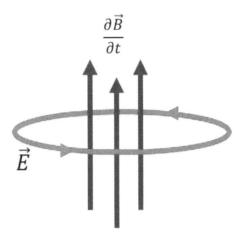

9.3 SELF INDUCTANCE

If you are familiar with the working of inductors, then you might know that Inductors have the ability to suppress variation in current flowing through it. An inductor is just a coil with some no of turns made of a conducting wire, so how do we explain this ability? The inductors ability to resist variation in current can be attributed to a phenomenon called Self Induction. The phenomenon can be better understood with the help of the figure below.

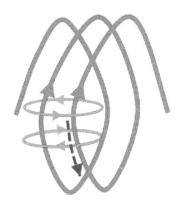

Consider just two turns of an inductor coil. When a current is passed through the inductor or more specifically the first turn of the inductor, it produces magnetic a field around it in a concentric manner (as with any other conductor). This magnetic field created by the first loop also links with the second loop, because of their proximity. The natural response of the second turn to this magnetic field, is to produce a current (or a counter magnetic field as represented by the bottom ring) such as to oppose the original current, in accordance with the Lenz's law. The direction of the current induced in the second turn due to the field generated by the first loop is show by the dotted arrow. These currents will be generated whenever there is a variation in current in the inductor and it opposes the original inductor current. This ability of an Inductor to oppose change in current is called the Self Inductance or simply Inductance.

Our analysis was just with 2 turns, but had there been more number of turns in the coil, the inductance will increase, because the magnetic field from one turn will have more turns to interact with. So, self-induction in a way is the mutual induction between the adjacent turns of the same coil.

So, when electromagnetic induction occurs between 2 coils, there are 3 inductances to consider; the self-inductance of coil 1, the self-inductance of coil 2 and the mutual inductance between the coils.

10. MAXWELL'S EQUATIONS

10.1 INTRODUCTION

In 1865, renowned physicist James Clerk Maxwell published his theory uniting electricity and magnetism with the help of a set of equations, which we now call as the Maxwell's equations. He first used the equations to propose that light is an electromagnetic phenomenon. Today these equations form the foundation of classical electromagnetism, optics, and various other fields of study.

As far as this book is concerned, this is the final piece of the Electromagnetism puzzle, everything we discussed so far is to get to this point. Understanding and interpreting the Maxwell's equations and their implications is one of "the goals" of Electromagnetic study. Let's push through this final bit.

Over the last 9 chapters, we derived the 4 Maxwell's equations as listed below.

Corresponding Law	Integral form	Point form
Gauss's Law for Electricity	$\oint_S \vec{D} \cdot \vec{ds} = \int_V \rho_v \, dv$	$\nabla \cdot \vec{D} = \rho_v$
Ampere's Law	$\oint_C \vec{H} \cdot \vec{dl} = \int_S \vec{J} \cdot \vec{ds}$	$\nabla \times \vec{H} = \vec{J}$
Gauss's Law for Magnetism	$\oint \vec{B} \cdot \vec{ds} = 0$	$\nabla \cdot \vec{B} = 0$
Faraday's Law	$\oint \vec{E} \cdot \vec{dl} = -\frac{d}{dt}(\int_S \vec{B} \cdot \vec{ds})$	$\nabla \times \vec{E} = -\frac{d(\vec{B})}{dt}$

We chose to express these equations in terms of E, D, B and H as a matter of convenience, but it's perfectly fine to express them using just E or D and B or H. By the way it was Maxwell himself that starting using the differential form for these equations.

Before moving ahead we'll quickly recap each of these laws one by one.

Gauss's law for Electricity:

The gauss's law for electricity states that "**The electric flux passing through any closed surface is equal to the total charge enclosed by that surface**". This law relates the electric field to the charge distribution that produced it.

Integral form	Point form
$\oint_S \vec{D} \cdot \vec{dS} = \int_V \rho_v \, dv$	$\nabla \cdot \vec{D} = \rho_v$

Gauss's law for Magnetism:

The gauss's law for magnetism states that *"The magnetic flux through a closed surface is always zero"*. This law essentially emphasizes on the fact that magnetic monopoles do not exist in nature. Magnets always exist as dipoles.

Integral form	Point form
$\oint \vec{B} \cdot \vec{dS} = 0$	$\nabla \cdot \vec{B} = 0$

Faraday's law:

According to the Faraday's laws, *"An EMF will be induced in a conductor that is placed in a varying magnetic field and this induced EMF is directly proportional to the rate of change of magnetic flux"*. The main takeaway from this law is that both electricity and magnetism aren't really 2 unrelated phenomena, they are 2 sides of the same coin.

Integral form	Point form
$\oint \vec{E}.\vec{dl} = -\frac{d}{dt}(\int \vec{B}.\vec{dS})$	$\nabla \times \vec{E} = -\frac{d(\vec{B})}{dt}$

Yes, we deliberately missed the Ampere's law. Maxwell's first great achievement was to club these laws together. His second big achievement was to identify the shortcoming of the Ampere's law.

Maxwell realized that the Ampere's law holds only for steady magnetic fields, but when it comes to varying magnetic fields, the law fails. So he took it on himself to modify the ampere's law and generalize it. The result is the Ampere-Maxwell law (discussed in the next section).

10.2 DISPLACEMENT CURRENT

So why is the Ampere's law not true in general? Let's use the help of some

math. According the Ampere's law,
$$\nabla \times \vec{H} = \vec{J}$$

Take the divergence on both sides of this equation
$$\therefore \nabla \cdot (\nabla \times \vec{H}) = \nabla \cdot \vec{J}$$

We know that the divergence of the curl of a vector is zero, so the LHS becomes zero.

$$\Rightarrow \nabla \cdot \vec{J} = 0$$

Now, from the continuity equation, we know

$$\nabla \cdot \vec{J} = \frac{-\partial \rho_v}{\partial t}$$

$$\Rightarrow \frac{\partial \rho_v}{\partial t} = 0$$

The continuity equation is based on the conservation of charge, so that holds no matter what. So if the divergence of J is zero, that means the charge density can never change or in other words, a charge can never enter or exit any closed surface. This is certainly not true except in case of static charges.

Now consider a practical case. If you are familiar with circuits, you might recall that an AC current can pass through a capacitor or that a capacitor acts like a short circuit when AC is applied. But we know that a capacitor is just 2 metal plates separated by a dielectric and therefore there is no way actual charges is transported between the plates. In this case, if you were to consider an amperian loop between the plates, there is a magnetic field, but there is no enclosed current. How is this possible?

Both the mathematical proof and the above practical example suggests that there is a charge-less current. Maxwell called it the Displacement current. To account for the displacement current, a new term J_d called the displacement current density was added to the Ampere's law.

$$\nabla \times \vec{H} = \vec{J} + \vec{J}_d$$

$$\boxed{\nabla \times \vec{H} = \vec{J} + \frac{\partial \vec{D}}{\partial t}}$$

In integral form, this equation can be written as,

$$\oint_C \vec{H} \cdot \vec{dl} = I_{enc} + \int_S \frac{\partial \vec{D}}{\partial t} \cdot \vec{dS}$$

This is the Ampere-Maxwell law.

10.3 ELECTROMAGNETIC WAVES

From the Faraday's law we know that a varying Magnetic field can produce an Electric field and from the Ampere-Maxwell law we know a varying Electric field can produce a Magnetic field. This tells us that the two phenomena are very much interlocked. Change in one generates another and vice versa. On the basis of this theory Maxwell had predicted the existence of something called electromagnetic waves.

These Electromagnetic waves consist of an oscillating Electric field and an oscillating Magnetic field. As the Electric field varies, a corresponding Magnetic field is generated. This Magnetic field in turn generates an Electric field and so on. This process is self-sustaining.

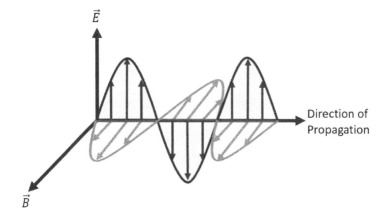

An electromagnetic wave is unlike a sound wave or any other kind of wave you see in nature and it has some interesting properties.

1. Both the Electric field and the Magnetic field will always be perpendicular to each other and to the direction of propagation. Hence, it's called a Transverse wave. In the figure above, the electric field is oscillating up and down and the magnetic field is oscillating towards you and away from you.
2. Both the Electric field and the Magnetic field will always be in phase i.e. they will have their

respective zeros and maximums at the same instant.
3. An Electromagnetic wave could propagate even without a medium, owing to the mutually assisting nature of the 2 fields.
4. Every charged particle in non-uniform motion produces an Electromagnetic wave.
5. Every point in an Electromagnetic wave satisfies the Maxwell's equations.

In his paper, Maxwell had not only predicted possibility of these waves, he even came up with a Wave equation to mathematically describe these waves. On top of that, he even calculated the speed at which these waves would propagate. Unfortunately, despite all the brilliant work from Maxwell, it took another 25 years before Heinrich Hertz first produced radio waves in his laboratory and rest as they say is history.

Using the Maxwell's equations and a few vector identities we can also derive the wave equation as follows:

From Faraday' law,

$$\nabla \times \vec{E} = -\frac{d(\vec{B})}{dt}$$

Taking curl on both sides of this equation,

$$\nabla \times (\nabla \times \vec{E}) = -\frac{d(\nabla \times \vec{B})}{dt} \quad \text{------ (1)}$$

Notice that we have interchanged the curl and the differential operator. This is permissible.

Now, using the vector identity $\nabla \times (\nabla \times \vec{A}) = \nabla(\nabla \cdot \vec{A}) - \nabla^2 \vec{A}$, we can rewrite equation (1) as,

$$\nabla(\nabla \cdot \vec{E}) - \nabla^2 \vec{E} = -\frac{d(\nabla \times \vec{B})}{dt} \quad \text{------ (2)}$$

But from Gauss's law for electricity, we know

$$\nabla \cdot \vec{E} = \frac{\rho_v}{\varepsilon}$$

But to make things easier, we'll assume a source free region ($\rho_v = 0$)

Therefore equation (2) becomes,

$$\nabla^2 \vec{E} = \frac{d(\nabla \times \vec{B})}{dt} \quad \text{------ (3)}$$

Using the Ampere-Maxwell law, we know,

$$\vec{\nabla} \times \vec{B} = \mu(\vec{J} + \varepsilon \frac{\partial \vec{E}}{\partial t})$$

We have to make one more assumption here, that it is a current free region i.e. J = 0.

Therefore equation (3) becomes,

$$\nabla^2 \vec{E} = \frac{\partial(\mu\varepsilon\frac{\partial \vec{E}}{\partial t})}{\partial t}$$

$$\boxed{\Rightarrow \nabla^2 \vec{E} = \mu\varepsilon \frac{\partial^2 \vec{E}}{\partial t^2}}$$

This is the vector wave equation describing nature of the Electric field in an EM wave. Instead of the Faraday's law, had we started with the Ampere-Maxwell law and proceeded in the same manner, we would have obtained this equation,

$$\boxed{\nabla^2 \vec{B} = \mu\varepsilon \frac{\partial^2 \vec{B}}{\partial t^2}}$$

This is the vector wave equation describing the nature of the Magnetic field in an EM wave.

The speed of propagation of these waves in free space is given by,

$$v = \frac{1}{\sqrt{\mu_0 \varepsilon_0}}$$

On calculating you'll get v = 3x 10^8 m/s. That's the speed of light! Maxwell was apparently shocked when he first calculated this result. In his own words, *"This velocity is so nearly that of light, that it seems we have strong reasons to conclude that light itself (including radiant heat, and other radiations if any) is an electromagnetic disturbance in the form of waves propagated through the electromagnetic field according to electromagnetic laws"*.

Anyway, that was the first time we knew that visible light is an Electromagnetic wave.

APPENDIX

1. Analogy between Electricity and Magnetism

Electricity	Magnetism
Charge	Current
Electric field Intensity (E)	Magnetic field Intensity (H)
Electric flux density (D)	Magnetic flux density (B)
Permittivity (ε)	Permeability (μ)
Gauss's law for Electricity ($\nabla \cdot \vec{D} = \rho_v$)	Gauss's law for Magnetism ($\nabla \cdot \vec{B} = 0$)
Faraday's law ($\nabla \times \vec{E} = -\frac{d(\vec{B})}{dt}$)	Ampere-Maxwell law ($\nabla \times \vec{H} = \vec{J} + \varepsilon \frac{\partial \vec{E}}{\partial t}$)
Polarization (P)	Magnetization (M)

2. Units

Quantity	Unit
Charge	coulomb (C)
Current	ampere (A)
Force	newton (N)
Permittivity	farad/meter (F/m)
Electric field Intensity	volt/meter (V/m)
Electric flux density	coulomb/meter^2 (C/m^2)
Potential	volt (V)
Current Density	ampere/meter^2 (A/m^2)
Conductivity	siemens/meter (S/m)
Electric Dipole moment	coulomb-meter (C-m)
Polarization	coulomb/meter^2 (C/m^2)
Permeability	henry/meter (H/m)
Magnetic field Intensity	ampere/meter (A/m)
Magnetic flux density	tesla (T) or weber/meter^2 (Wb/m^2)
Magnetization	ampere/meter^2 (A/m^2)
Inductance	henry
Magnetic moment	ampere-meter^2 (A-m^2)